THE SECOND DIGITAL TURN

Writing **Architecture** series

A project of the Anyone Corporation; Cynthia Davidson, editor

Earth Moves: The Furnishing of Territories
Bernard Cache, 1995

Architecture as Metaphor: Language, Number, Money
Kojin Karatani, 1995

Differences: Topographies of Contemporary Architecture
Ignasi de Solà-Morales, 1996

Constructions
John Rajchman, 1997

Such Places as Memory: Poems 1953–1996
John Hejduk, 1998

Welcome to The Hotel Architecture
Roger Connah, 1998

Fire and Memory: On Architecture and Energy
Luis Fernández-Galiano, 2000

A Landscape of Events
Paul Virilio, 2000

Architecture from the Outside: Essays on Virtual and Real Space
Elizabeth Grosz, 2001

Public Intimacy: Architecture and the Visual Arts
Giuliana Bruno, 2007

Strange Details
Michael Cadwell, 2007

Histories of the Immediate Present: Inventing Architectural Modernism
Anthony Vidler, 2008

Drawing for Architecture
Léon Krier, 2009

Architecture's Desire: Reading the Late Avant-Garde
K. Michael Hays, 2009

The Possibility of an Absolute Architecture
Pier Vittorio Aureli, 2011

The Alphabet and the Algorithm
Mario Carpo, 2011

Oblique Drawing: A History of Anti-Perspective
Massimo Scolari, 2012

A Topology of Everyday Constellations
Georges Teyssot, 2013

Project of Crisis: Manfredo Tafuri and Contemporary Architecture
Marco Biraghi, 2013

A Question of Qualities: Essays in Architecture
Jeffrey Kipnis, 2013

Noah's Ark: Essays on Architecture
Hubert Damisch, 2016

The Second Digital Turn: Design Beyond Intelligence
Mario Carpo, 2017

THE SECOND DIGITAL TURN

DESIGN BEYOND INTELLIGENCE

MARIO CARPO

THE MIT PRESS

CAMBRIDGE, MASSACHUSETTS

LONDON, ENGLAND

This book was set in Filosofia OT and Trade Gothic LT Std by Toppan Best-set Premedia Limited. Printed and bound in the United States of America.

Library of Congress Cataloging-in-Publication Data

Names: Carpo, Mario, author.
Title: The second digital turn : design beyond intelligence / Mario Carpo.
Description: Cambridge, MA : The MIT Press, 2017. | Series: Writing architecture | Includes bibliographical references and index.
Identifiers: LCCN 2016054313 | ISBN 9780262534024 (pbk. : alk. paper)
Subjects: LCSH: Architecture and technology. | Architecture--Information technology. | Architecture--Computer-aided design.
Classification: LCC NA2543.T43 C37 2017 | DDC 720.72--dc23 LC record available at https://lccn.loc.gov/2016054313

10 9 8 7 6 5 4 3 2 1

CONTENTS

ACKNOWLEDGMENTS

While researching and writing this book I had to dabble in an inordinate number of disciplines and subjects, including some that are manifestly outside of my expertise. I am aware of the risks this entails; specialists in each of those fields will no doubt find errors of all sorts. As often happens, I could only outline a more general picture to the detriment of local detail; going against the logic of the artificial intelligence I try to describe, I was often obliged to merge, neglect, or compress plenty of data in order to allow some visible patterns to emerge. I am grateful in advance to the scholars and colleagues who will correct my arguments and flag my simplifications and omissions. I am also thankful to the many colleagues and friends with whom I discussed the ideas in this book over the course of the last three years, and who generously offered tips and advice: in particular, Alisa Andrasek, Marjan Colletti, Marcos Cruz, Christian Girard, Jeff Huang, Achim Menges, Marco Panza, Gilles Retsin, Jenny Sabin, Patrik Schumacher, Axel Sowa, and the faculty and students at the B-Pro program at the Bartlett School of Architecture, with whom I had many fruitful sessions and discussions. Almost weekly discussions with Frédéric Migayrou left an evident trace throughout chapter 2, and Philippe Morel generously shared technical and mathematical insights, particularly on the history of spline making. A grant from the Bartlett School of Architecture allowed me to purchase some reproduction rights, and to hire Alexandra Vougia as a research assistant and Tina Di Carlo as a copy editor during the first phase of writing. Cynthia Davidson guided all stages of the making of the book, from conception and development to editing and delivery, with her usual flair and professionalism.

London, September 2016

1 INTRODUCTION

Architects tend to be late in embracing technological change. This chronic belatedness started at the very beginning of the Western architectural tradition: Vitruvius's *De Architectura*, one of the most influential books of all time, was composed in the early years of the Roman Empire, but it described a building technology that, by the time Vitruvius put it into writing, was already a few centuries old. Vitruvius refers for the most part to trabeated, post-and-lintel structures, and he doesn't even mention arches or vaults, which were already a major achievement of Roman engineering. When Vitruvius mentions bricks, he seems to have in mind the primitive sun-dried brick of the early Mediterranean and Mesopotamian traditions; yet when writing his treatise—as a retired military engineer living on a pension from the Roman army and a grant from the emperor's sister—he was probably sitting in a modern Roman house made of solid bricks baked in a furnace. Why did Vitruvius choose to celebrate an obsolete way of building, and concoct the fiendish plan to bequeath to posterity a building technology that nobody, at the time of his writing, was using any more? We don't know. But perhaps it should come as no surprise that his treatise soon fell into oblivion, only to be revived fifteen centuries later by the Humanists of the Italian Renaissance, who, of course, could not make heads or tails of Vitruvius's often opaque technological and scientific lore. The most alert among Vitruvius's Renaissance readers did remark, meekly, that Vitruvius's treatise, and the extant Roman ruins they could still peruse all over Italy, did not seem to match.

But again, the technological ambitions of most Renaissance architects were simple at best, and early modern classicists did not need much technology to build in the classical styles they cherished. When the Renaissance came, European architecture had just gone through an age of astounding technical renewal: the high rises of Gothic spires and pinnacles were so daring and original that we still do not know how they were built (and we would struggle to rebuild them if we had to use the tools and materials of their time). But when Renaissance classicists and their Italianate style took over, the technical skills of the medieval master builders were abandoned, and early modern architecture fell back on the good old post-and-lintel structures of classical antiquity, this time with arches, vaults, and domes added when needed.

For centuries, and with few exceptions, modern classicism continued to stifle technological innovation in building: in the nineteenth century, while the Industrial Revolution was changing society, the world, and the way we build, architects mostly used the new industrial materials to imitate the shapes and styles of classical antiquity (and, at times, of other historical periods too). Even the golden age of twentieth-century modernism, when architects finally decided to come to terms with the industrial world, was—when all the pizzazz is taken away—a sadly *retardataire* phenomenon. Look at the makers of cars, planes, or steamships, Le Corbusier said in his famous writings of the early 1920s: unlike us, they know how to deal with today's technologies of mass production and how to exploit the assembly line; we should take our lead from them, he concluded, and imitate their example. Taylorism and Fordism were not invented by architects; architects just followed suit—or tried to: a controversial, painful, and often not very successful travail. For houses, unlike automobiles or washing machines, can hardly be identically mass-produced: to this day, with few exceptions, that is still technically impossible. Besides, some always thought that standardized housing was never such a good idea to begin with. For late twentieth-century postmodernists, for example, every human dwelling should be a one-off, a unique work of art, made to measure and made to order, like a bespoke suit. Of course, bespoke suits are expensive, but many noted design professionals never objected to that either.

Some, however, did and do. This may explain why, when the digital turn came in the 1990s, architects—not all of them, but the best—adopted digital tools and embraced digital change sooner than any other trade, industry, or creative profession. For this was a technology meant to produce variations, not identical copies; customized, not standardized products: far more than a postmodern dream come true, variability is a deep-rooted ambition of architects and designers, craftsmen and engineers of all times and places. Architects did not invent digital design and fabrication: numerically controlled milling machines had been around since the late 1940s, and affordable CAD software since the start of the PC revolution (circa 1982); economists and technologists started discussing mass customization in the late 1980s. But most of the early discussions on product differentiation focused on small-batch production, low-volume manufacturing, and multiple-choice marketing strategies. These were ways to mitigate the scale of standardized mass production, but without abandoning its technical logic: the more copies we make, the cheaper each copy will be; products made in smaller series may be better targeted to customer needs, but will be more expensive. In the 1990s, to the contrary, the first generation of digitally intelligent designers had a simple and drastic idea. Digital design and fabrication, they claimed, should not be used to emulate mechanical mass production but to do something else—something that industrial assembly lines cannot do.

Digital fabrication does not use mechanical matrixes, casts, stamps, molds, or dies, hence it does not need to reuse them to amortize their cost. As a result, making more digital copies of the same item will not make any of them any cheaper, or, the other way around, each digitally fabricated item can be different, when needed, at no additional cost: the mass production of variations is the general mode—the default mode, so to speak—of a digital design and fabrication workflow. Digital mass customization is one of the most important ideas ever invented by the design professions: an idea that is going to change, and to some extent has already changed, the way we design, produce, and consume almost everything, and one that will subvert—and to some extent has already subverted—the cultural and technical foundations of our civilization. And for better or worse, digital mass customization was *our* idea: it was developed, honed, tested, and conceptualized in a handful of schools of architecture in Europe and the United States in the 1990s. To this day, designers and architects are the best specialists in it: designers and architects—not technologists or engineers, not sociologists or philosophers, not economists or bankers, and certainly not politicians, who still have no clue about what is going on.

It would take a great historian, mathematician, and philosopher to explain how and why this epoch-making cultural and technical revolution spawned a new architectural style based on smooth and curving lines and surfaces. There is evidence that none of the early protagonists of the digital turn in architecture—first and foremost Peter Eisenman—ever anticipated that. Yet the style of the blob, also known as the style of the spline or of digital streamlining, became the hallmark of the first digital age in the 1990s. Today both trends—the technical one and the stylistic one—often go under the rubric of parametricism; back then both fell on hard times when, early in the new millennium, the Internet bubble burst. With the collapse of the new "digital economy," the wave of digital exuberance and technological optimism of the late 1990s suddenly lost traction, and many in the design professions started to lambast the digital blob as the most conspicuous symbol of an age of excess, waste, and technological delusion.

When the dust settled, a new spirit and some new technologies, which could exploit the infrastructural overinvestment of the 1990s at a discount, led to what was then called the Web 2.0: the participatory Web, based on collaboration, interactivity, crowdsourcing, and user-generated content. But the ensuing meteoric rise of social media, from Facebook to *Wikipedia*, was not matched by any comparable development in digital design. In fact, at the time of this writing (2016) it seems safe to conclude that the much touted and much anticipated shift from mass customization to mass collaboration has not happened. With the exception of a handful of avant-garde experiments, and more remarkably of a family of technologies known as Building Information Modeling, or BIM—unanimously adopted by the building and construction industry but reviled by the trendiest creatives and in academia—the design professions seem to have flatly rejected a techno-cultural development that would weaken (or, in fact, recast) some of their traditional authorial privileges.

At the same time, many of the ideas that the digital avant-garde came up with and test-drove in the course of the 1990s now seem to have taken on a life of their own, spreading like wildfire in all spheres of today's society, economy, and culture. The principles of digital mass customization, and to some extent of collaborative design, have moved from the manufacturing of physical objects (teapots, chairs, buildings) to the creation and consumption of media objects (text, images, music), and lastly to the production of immaterial objects, such as contracts and agreements bearing on all kinds of legal and financial transactions: pricing, rentals, employment, services, the supply and trade of electricity, and the issuance and circulation of debt (which includes the creation of money). Just as in the 1990s we discovered that digital mass customization can deliver economies of production without the need for scale, today we are learning that the aggregation of supply and demand does not make digitally mass-customized transactions any cheaper: the cost of processing most transactions in a nonstandard, algorithmic environment is unaffected by size. Transactions bearing on items of negligible or irrelevant import used to be too expensive, unwieldy, forbidden by law, or collectively regulated by charters or statutes. But today, one-to-one bespoke contracts of any import and size can be practically implemented using digital tools. As a result, many regulations and regulators that standardized transactions and transacted items in the pursuit of scale are now technically unwarranted. Some of such traditional regulators are fast becoming culturally and politically irrelevant too: the modern nation-state, which was indispensable to achieving economies of scale during the Industrial Revolution, is a case in point.

I discuss some of these issues, and the role of the digital avant-garde in the invention of this new techno-social paradigm, in the second part of this book (chapter 4, "The Participatory Turn That Never Was," and chapter 5, "Economies Without Scale: Toward a Nonstandard Society"). Yet, while all of this may well point to our collective societal future, in digital design and fabrication most of this is already history—albeit a recent one. Digitally intelligent designers may well have invented, or at least intuited, the core principles of the first digital turn one generation ago. But then something else came up, and the digital avant-garde again took notice. To make a long story short, at some point early in the new millennium some digital tools started to function in a new way, as if following a new and apparently inscrutable logic—the "search, don't sort" logic of the new science of data. The theoretical implications of this new technical paradigm were not clear from the start. Regardless, for the last ten years or so digitally intelligent designers have been busy coping and dealing with these new processes, trying to compose with them and putting them to task. That is the subject of the first part of this book (chapter 2, "The Second Digital Turn," and chapter 3, "The End of the Projected Image").

Twenty to thirty years is a long time in the annals of information technology—long enough to allow us to discern a fundamental rift between the inner workings of yesterday's and today's computational tools. At the beginning, in the 1990s, we used our brand-new digital machines to implement the old science we knew—in a sense, we carried all the science we had over to the new computational platforms we were then just discovering. Now, to the contrary, we are learning that computers can work better and faster when we let them follow a different, nonhuman, postscientific method; and we increasingly find it easier to let computers solve problems *in their own way*—even when we do not understand what they do or how they do it. In a metaphorical sense, computers are now developing their own science—a new kind of science. Thus, just as the digital revolution of the 1990s (new machines, same old science) begot a new way of making, today's computational revolution (same machines, but a brand new science) is begetting a new way of thinking.

Evidently the idea that inorganic machines may nurture their own scientific method—their own intelligence, some would say—lends itself to various apocalyptic or animistic prophecies. This book follows a different, more arduous path. Designers are neither philosophers nor theologians. They may be prey to beliefs or ideologies, but no more and no less than in most other professions. By definition, designers make real stuff, hence they are bound to some degree of philistinism: they are paid only when the stuff they make works—or when they can persuade their clients that at some point it will. And based on the immediate feedback we get in the ordinary practice of our trade, it already appears that, to chart the hitherto untrodden wilds of posthuman intelligence, some strategies work better than others. Having humans imitate computers does not seem any smarter than having computers imitate humans. *À chacun son métier*: to each its trade.

Ultimately, the task of the design professions is to give shape to the objects we make and to the environment we inhabit. In the 1990s we invented and interpreted a new cultural and technical paradigm; we were also remarkably successful in creating a visual style that defined an epoch and shaped technological change. It is too soon to tell if we will carry it off again this time around; the second digital turn has just started, and the second digital style is still in the air. We may have the best ideas in the world—and I wrote this book precisely because it seems to me that digitally intelligent designers are finding and testing capital new ideas right now: just like in the 1990s, well ahead of anyone else. Yet, in the end, no one will take us seriously unless the stuff we make looks good.

2 THE SECOND DIGITAL TURN

The collection, transmission, and processing of data have been laborious and expensive operations since the beginning of civilization. Writing, print, and other media technologies have made information more easily available, more reliable, and cheaper over time. Yet, until a few years ago, the culture and economics of data were strangled by a permanent, timeless, and apparently inevitable scarcity of supply: we always needed more data than we had. Today, for the first time, we seem to have more data than we need. So much so, that often we do not know what to do with them, and we struggle to come to terms with our unprecedented, unexpected, and almost miraculous data opulence. As always, institutions, corporations, and societies, the cultural inertia of which seems to grow more than proportionally to the number of active members, have been slow to adapt. Individuals, on the contrary, never had much choice. Most Westerners of my generation (the last of the baby boomers) were brought up in the terminal days of a centuries-old small-data environment. They laboriously learned to cope with its constraints and to manage the endless tools, tricks, and trades developed over time to make the best of the scant data they had. Then, all of a sudden, this data-poor environment just crumbled and fell apart—a fall as unexpected as that of the Berlin Wall, and almost coeval with it. As of the early 1990s, digital technologies introduced a new culture and a new economics of data that have already changed most of our ways of making, and are now poised to change most of our ways of thinking.

My first memorable clash with oversized data came, significantly, in or around May 1968, and it was due to an accident unexplained to this day. I was then in primary school, and on a Wednesday afternoon our teacher sent us home with what appeared from the start to be a quirky homework assignment: a single division problem, but between two very big numbers. As we had no school on Thursdays (a tradition that preceded the modern trend for longer weekends), homework on Wednesday tended to be high-octane, so as to keep us busy for one full day. That one did. Of the two numbers we had to tackle, the dividend struck the eye first, as it appeared monstrously out of scale; the divisor was probably just three or four digits, but this is where our computational ordeal started. It soon turned out that the iterative manual procedure we knew to perform the operation, where the division was computed using Hindu-Arabic integers on what was in fact a virtual abacus drawn on paper, became very unwieldy in the case of divisors larger than a couple of digits.

This method, I learned much later, was more or less still the same one that Luca Pacioli first set forth in 1494.[1] I do not doubt that early modern abacists would have known how to run it with numbers in any format, but we didn't; besides, I have reason to suspect that our divisor might have been, perversely, a prime number (but as we did not know fractions in fourth grade, that would not have made any difference). So on Thursday morning, after some perplexity, I tried to circumvent the issue with some leaps of creative reckoning, and as none came to any good, during lunch break—which at the time still implied a full meal at home for everyone working in town—I threw in the towel, and asked my father. He looked at the numbers with even more bemused perplexity, mumbled something I was not meant to hear, and told me to call him back in his office later in the afternoon. Did he not have a miraculous instrument in his breast pocket, a slide rule that I had seen him use to calculate all sorts of stuff, including a forecast for a soccer match? It would be of no use in this instance, my father answered, because using the slide rule he could only get to an approximate result, and my teacher evidently expected a real number—all digits of it, and a leftover to boot.

So I waited and called his office in the afternoon. I dictated the numbers and I heard them punched into an Olivetti electromechanical Divisumma calculator, at the time a fixture on most office desks in Europe. I knew those machines well—I often played with them when the secretaries were not there. Under their hood, beautifully streamlined by Marcello Nizzoli, a panoply of levers, rods, gears, and cogwheels noisily performed additions, subtractions, multiplications, and, on the latest models, even divisions. But divisions remained the trickiest task: the machine had to work on them at length, and the bigger the numbers, the longer the time and labor needed for number crunching. After some minutes of loud clanging there was a magic hiatus when the machine appeared to stand still, and then a bell rang, and the result was printed in black and red ink on a paper ticker. That day, however, that liberating sound never came, as the dividend in my homework was, I was told, a few digits longer than the machine could take. I could get an approximate result, which once again was likely not what I should bring to school on Friday morning. Then my father stepped in again: are you not at school with young X, the son of the bank director, he asked—call him, for they likely have better machines over there. And so I did, and young X told me he had indeed called his father, and he was waiting to hear back. I thought then, as I do now, that he was lying.

Early on Friday morning, while waiting in front of the school gates, some of us tried to compare results. Among those who had some to show, all results were widely different. Young X gloated and giggled and did not participate in the discussion. He went into politics in the 1980s, and to jail in the 1990s, one of the first local victims of the now famous "Mani Pulite" (clean hands) judicial investigation. Back then, however, when the bell rang and the teacher came in to the class, all we wanted to know was the right result. The teacher stood up from his desk and looked around, somewhat ruffled and dazzled, holding a batch of handwritten notes. Then, before he could utter a word, he fainted in front of us all, and fell to the floor. Medics were called, and we were sent to the courtyard. When we were let back in a few hours later, an old retired teacher, hastily recalled, told us jokes and stories for the rest of the day. We finished the school year with a substitute teacher; our titular teacher never came back, and nobody knows what his lot was after that day. There were unconfirmed rumors in town that he had started a new life in Argentina. And to this day, I cannot figure out why on his last day as a schoolteacher he would give us an assignment that evidently outstripped our, and probably his own, arithmetical skills—but also far exceeded the practical computational limits of all tools we could have used for that task. Hindu-Arabic integers still worked well around 1968, precisely because nobody tried to use them to tackle such unlikely magnitudes, which seldom occurred in daily life, or, indeed, in most technical or financial trades.

Hindu-Arabic numerals were a major leap forward for Europe when they were adopted (first, by merchants) in the fifteenth century. Operations with Hindu-Arabic numerals—then called *algorism*, from the Latinized name of the Baghdad-based scientist who first imported them from India to the Arabic world early in the ninth century[2]—worked so much better than all other tools for quantification then in use in Europe that they would soon replace them all (Latin numerals were the first to go, but algebra and calculus would soon phase out entire swaths of Euclidian geometry, too). Number-based precision was a major factor in the scientific revolution, and early modern scientists were in turn so successful in their pursuit of precision that they soon outgrew the computational power of the Hindu-Arabic integers at their disposal. Microscopes and telescopes, in particular, opened the door to a world of very big and very small numbers, which, as I learned on that memorable day in May 1968, could present insurmountable barriers to number-based, manual reckoning. Thus early modern algorism soon went through two major upgrades, first with invention of the decimal point (1585), then of logarithms (1614).[3]

A masterpiece of mathematical ingenuity, logarithms are one of the most effective data-compression technologies of all time. By translating big numbers into small numbers, and, crucially, by converting the multiplication (or division) of two big numbers into the addition (or subtraction) of two small numbers, they made many previously impervious arithmetical calculations much faster and less error-prone. Laplace, Napoleon's favorite mathematician, famously said that logarithms, by "reducing to a few hours the labor of many months, doubled the life of the astronomer."[4] Laplace went on to claim that, alongside the practical advantages they offer, logarithms are even more praiseworthy for being a pure invention of the mathematical mind, unfettered by any material or manual craft; but in that panegyric Laplace was curiously blind to a crucial, indeed critical, technical detail: logarithms can serve some practical purpose only when paired with logarithmic tables, where the conversion of integers into logarithms, and the reverse, are laboriously precalculated. Logarithmic tables, in turn, would be useless unless printed. If each column of a logarithmic table—an apparently meaningless list of millions of minutely scripted decimal numbers—had to be copied by hand, the result would be too labor-intensive to be affordable and too error-prone to be reliable. Besides, errors would inevitably accrue over time, from one copy to the next, as in all chains of scribal transmission.[5]

The efficacy of logarithms is thus dependent on what today we would call economies of scale in supply-chain management: many repetitive calculations that no end user would perform in person are ingeniously centralized, executed once and for all by a team of specialists, double-checked, then distributed in ready-made form to each customer through print. By copying numbers from the printed tables, each user acquires in fact a prefabricated, mass-produced modular item of calculation, which he imports and incorporates into his own computational work. Economies of scale are proportional to the number of identical items produced: in this instance, the profit for all rises if the same tables can be printed and handed out to many customers. Failing that, logarithms would not have any reason to exist—except as a mathematical oddity, much as the binary notation was before the invention of electric computers.[6] Logarithms are the quintessential mathematics of the age of printing; a data-compression technology of the mechanical age.

As it happens, my own encounter with logarithms in high school, circa 1973, coincided with my first acquaintance with a handheld electronic calculator, which was probably a 1971 Brother ProCal. Although it was, in theory, possible to hold it in hand, that machine was the size of a brick and almost as heavy; it did nothing beyond the four operations and, the display being limited to eight digits, it could not compete with the logarithmic tables I was then learning to use. As a result, I dutifully kept practicing with logarithms for all of the three or so months that our high school curriculum still devoted to that baroque (literally) cultural technology.[7] I did not know then that only three or four years later, during my second year in college, I would buy (at a price then approximately equal to that of ten pizzas) a Texas Instruments TI-30, which included trigonometric, exponential,

Chilias vicesima.

Num. absolut.	Logarithmi.	Num. absolut.	Logarithmi.	Num. absolut.	Logarithmi.
19801	4,29668,71237,7240 2,19324,0320	19834	4,29741,03088,9739 2,18959,1278	19867	4,29813,22917,7961 2,18595,4357
19802	4,29670,90561,7560 2,19312,9564	19835	4,29743,22048,1017 2,18948,0890	19868	4,29815,41513,2318 2,18584,4336
19803	4,29673,09874,7124 2,19301,8820	19836	4,29745,40996,1907 2,18937,0514	19869	4,29817,60097,6654 2,18573,4327
19804	4,29675,29176,5944 2,19290,8087	19837	4,29747,59933,2421 2,18926,0149	19870	4,29819,78671,0981 2,18562,4327
19805	4,29677,48467,4031 2,19279,7364	19838	4,29749,78859,2570 2,18914,9794	19871	4,29821,97233,5308 2,18551,4340
19806	4,29679,67747,1395 2,19268,6654	19839	4,29751,97774,2364 2,18903,9452	19872	4,29824,15784,9648 2,18540,4363
19807	4,29681,87015,8049 2,19257,5954	19840	4,29754,16678,1816 2,18892,9120	19873	4,29826,34325,4011 2,18529,4397
19808	4,29684,06273,4003 2,19246,5265	19841	4,29756,35571,0936 2,18881,8799	19874	4,29828,52854,8408 2,18518,4442
19809	4,29686,25519,9268 2,19235,4587	19842	4,29758,54452,9735 2,18870,8489	19875	4,29830,71373,2850 2,18507,4499
19810	4,29688,44755,3855 2,19224,3921	19843	4,29760,73323,8224 2,18859,8191	19876	4,29832,89880,7349 2,18496,4566
19811	4,29690,63979,7776 2,19213,3266	19844	4,29762,92183,6415 2,18848,7903	19877	4,29835,08377,1915 2,18485,4645
19812	4,29692,83193,1042 2,19202,2622	19845	4,29765,11032,4318 2,18837,7627	19878	4,29837,26862,6560 2,18474,4734
19813	4,29695,02395,3664 2,19191,1989	19846	4,29767,29870,1945 2,18826,7362	19879	4,29839,45337,1294 2,18463,4835
19814	4,29697,21586,5653 2,19180,1368	19847	4,29769,48696,9307 2,18815,7108	19880	4,29841,63800,6129 2,18452,4947
19815	4,29699,40766,7021 2,19169,0757	19848	4,29771,67512,6415 2,18804,6865	19881	4,29843,82253,1076 2,18441,5069
19816	4,29701,59935,7778 2,19158,0158	19849	4,29773,86317,3280 2,18793,6633	19882	4,29846,00694,6145 2,18430,5203
19817	4,29703,79093,7936 2,19146,9570	19850	4,29776,05110,9913 2,18782,6412	19883	4,29848,19125,1348 2,18419,5348
19818	4,29705,98240,7506 2,19135,8993	19851	4,29778,23893,6325 2,18771,6203	19884	4,29850,37544,6696 2,18408,5504
19819	4,29708,17376,6499 2,19124,8427	19852	4,29780,42665,2528 2,18760,6004	19885	4,29852,55953,2200 2,18397,5671
19820	4,29710,36501,4926 2,19113,7872	19853	4,29782,61425,8532 2,18749,5817	19886	4,29854,74350,7871 2,18386,5849
19821	4,29712,55615,2798 2,19102,7329	19854	4,29784,80175,4349 2,18738,5641	19887	4,29856,92737,3720 2,18375,6037
19822	4,29714,74718,0127 2,19091,6796	19855	4,29786,98913,9990 2,18727,5475	19888	4,29859,11112,9757 2,18364,6238
19823	4,29716,93809,6923 2,19080,6276	19856	4,29789,17641,5465 2,18716,5321	19889	4,29861,29477,5995 2,18353,6449
19824	4,29719,12890,3199 2,19069,5765	19857	4,29791,36358,0786 2,18705,5178	19890	4,29863,47831,2444 2,18342,6671
19825	4,29721,31959,8964 2,19058,5266	19858	4,29793,55063,5964 2,18694,5046	19891	4,29865,66173,9115 2,18331,6904
19826	4,29723,51018,4230 2,19047,4779	19859	4,29795,73758,1010 2,18683,4926	19892	4,29867,84505,6019 2,18320,7148
19827	4,29725,70065,9009 2,19036,4302	19860	4,29797,92441,5936 2,18672,4816	19893	4,29870,02826,3167 2,18309,7404
19828	4,29727,89102,3311 2,19025,3837	19861	4,29800,11114,0752 2,18661,4717	19894	4,29872,21136,0571 2,18298,7670
19829	4,29730,08127,7148 2,19014,3382	19862	4,29802,29775,5469 2,18650,4629	19895	4,29874,39434,8241 2,18287,7947
19830	4,29732,27142,0530 2,19003,1939	19863	4,29804,48426,0098 2,18639,4553	19896	4,29876,57722,6188 2,18276,8236
19831	4,29734,46145,3469 2,18992,1507	19864	4,29806,67065,4651 2,18628,4487	19897	4,29878,75999,4424 2,18265,8535
19832	4,29736,65137,5976 2,18981,2086	19865	4,29808,85693,9138 2,18617,4433	19898	4,29880,94265,2959 2,18254,8845
19833	4,29738,84118,8062 2,18970,2677	19866	4,29811,04311,3571 2,18606,4390	19899	4,29883,12520,1804 2,18243,9167
19834	4,29741,03088,9739 2,18959,1278	19867	4,29813,22917,7961 2,18595,4357	19900	4,29885,30764,0971 2,18232,9499

19901

2.1

Henry Briggs, *Arithmetica Logarithmica, sive Logarithmorum chiliades triginta* (London: William Jones, 1624), one page from the tables.

and logarithmic functions. That miraculous machine wiped out, overnight, most of the chores that had encumbered our math and engineering classes.

Just like divisions in the old electromechanical Olivettis, the exponential and logarithmic functions stretched the performance of that calculator to its limits: most operations were calculated instantaneously, but sometimes, when dealing with very large numbers, the machine hesitated: there was a momentary lag, and the red display blinked for a few instants before the result was shown—a delay we generously interpreted as a machinic analog, and almost an empathic reenactment, of the suffering and effort intrinsic to human computing. That quirk aside, the calculator worked magnificently, and the first moment of cultural awareness of a digital revolution in the making—the first *ceci tuera cela* epiphany: in this case, of the computer killing the book—came when we realized that we would no longer need any trigonometric or logarithmic tables in print, as all logarithms, for example, could be instantaneously calculated by the machine at all times.

But that was only the start, as we soon realized that the logarithm function itself must have been put on the machine's keyboard as a practical joke, or a trap for fools. Why would anyone use logarithms to convert big numbers into small numbers before feeding those numbers back into the machine, when the same machine could work with any numbers of any size almost at the same speed? Was there any evidence that processing fewer big numbers, rather than many smaller ones, would wear out the machine, damage it, or drain its 9V battery? Did the machine have a limited computational life span, and die after performing a given number of operations? That was certainly the case for the electromechanical Olivettis: their cogs and rods did wear out, and had to be kept greased, maintained, and replaced after a certain number of cycles, like all mechanical tools. But there were no moving parts in the solid-state TI-30, nor did its processor make any noise when at work. Aside from the blinking screen, which indicated some degree of machine stress in certain conditions, one was inclined to conclude that, by and large, using the machine in full would not cost more than using it sparingly. If so, the use of logarithms and antilogarithms to compress and decompress numbers before and after use, so to speak, would only add time and errors. Logarithms, as Laplace had pointed out, had indeed extended the life of astronomers—and, we could add today, of many nineteenth- and twentieth-century engineers; but do logarithms extend the life of a solid-state processor? Given the power of even the cheapest of today's computing tools, the cost of electronic number crunching is, in most cases, not significantly affected by relatively small variations in the size of the numbers we type in, and for most applications occurring at normal technical scales there is no need to laboriously process the numbers we use in order to shave a few digits off any of them. Not surprisingly, today's engineers, and even astronomers, no longer use logarithms. Computers have de-invented logarithms; logarithms are a technology for data compression that digital computers simply do not need any more.

Today's computers are not, in essence, different from those of 1978: they still use binary numbers and electric impulses to process them. But they are far more powerful, faster, and cheaper due to steady technical progress in computing hardware. Moore's Law, first formulated in 1965, noted that the numbers of transistors per square inch of an integrated circuit doubles every two years, a growth rate that has been more or less maintained to this day. Most measures of speed, capacity, and even price in electronic computing, whether directly determined by the performance of a silicon chip or not, have moved on a similar scale. Yet this oddly regular pace of quantitative advancement seems to have recently crossed some crucial threshold, as proven by the popular currency and universal appeal of the idea of Big Data, which has been mooted in all kinds of contexts and peddled for all kinds of purposes since the late 2000s.

The term "Big Data" originally referred simply to our technical capacity to collect, store, and process increasing amounts of data at decreasing costs, and this meaning still stands, regardless of hype, media improprieties, or semantic confusion.[8] There is nothing inherently revolutionary in this idea, which per se could equally refer to the advantages of writing over orality, of print over scribal transmission, or to each incremental technical improvement of digital technologies for at least the last fifty years. What is new, however, and specific to our time, is the widespread and growing realization that today's economy of data may be unprecedented in history. This is what has brought Big Data front and center; and there may be more to this than media spin. For this cultural perception may well be the main divide between the first digital revolution, in the 1990s, and the second, in the 2010s. In the 1990s the old logic of small data, laboriously crafted through so many centuries of limited data supply, still applied to computing in full. Computers were already processing data faster and more effectively than any other technology in history, but data were still seen as expensive ware, to be used as parsimoniously as possible. Now, for the first time ever, data are seen as abundant and cheap, and, more important, as becoming more abundant and cheaper every day.

If this trend continues, one may reasonably project the evolution of this cost curve asymptotically into the future and conclude that, at some point, an almost infinite amount of data will be recorded, transmitted, and retrieved at almost no cost. Evidently, a state of zero-cost recording and retrieval will always be impossible; yet this is the tendency as we perceive it—this is where today's technology seems to be heading. If this is so, then we must also come to the inevitable conclusion that many technologies of data compression still in use will at some point become unnecessary, as the cost of compressing and decompressing the data (at times losing some in the process) will be greater than the cost of keeping all raw data in their pristine state for a very long time, or even forever. If we say digital data-compression technologies, we immediately think of JPEG or MP3. But as the example of logarithms suggests, many cultural technologies that humankind has developed over time, from the alphabet to the most sophisticated information retrieval systems, should be seen as primarily, or accidentally, developed in order to cope with what was, until recently, an inevitable material condition of data penury affecting all people at all times and in all places—a chronic shortage of data, which today, all of a sudden, has just ceased to be.

2.1 Data-Compression Technologies We Don't Need Anymore

The list of cultural technologies being made obsolete by today's new data affluence is already a long one. To some extent, all of the most successful cultural technologies in history must have been at their start, either by design or by chance, data-compression technologies. As information processing was always expensive and technically challenging, only media that could offer good data-compression rates—that is, that could trim the size of messages without losing too much content—could find a market, take root, and thrive. The alphabet, an old and effective compression technology for sound recording, is a case in point.

The alphabetic notation converts the infinite variations of sounds produced by the human voice (and sometimes a few other sounds and noises too) into a limited number of signs, which can be easily recorded and transmitted across space and time.[9] Anyone trained to literacy in a given alphabetical language can record the sound of someone else's voice through writing (taking dictation, for example), as well as reproduce a sound that was notated by someone else without having heard the voice of the original speaker, and regardless of meaning—even when that sound has no meaning at all (supercalifragilisticexpialidocious, for example). This strategy worked well for centuries, and it still allows us to read transcripts from the living voices of famous people we never listened to and who never wrote a line, such as Homer, Socrates, or Jesus Christ. In time, the alphabet was adapted to silent writing and silent reading, and its original function as a voice recorder was lost, but its advantages as a data-compression technology were not. The Latin alphabet, for example, records all sounds of the human voice, hence most conveyable thoughts of the human mind, using a combination of less than thirty signs. At some point these signs began to be mass-produced in metal, which allowed for lines of type, and entire books, to be reproduced from mechanical matrixes that in turn were made from the combination of a very limited number of interchangeable parts: the alphabet was the software that made the hardware of print from moveable type possible.[10] Moveable type could be reproduced as standard metal type pieces precisely because the pieces were limited in number, thus allowing for economies of scale in the mass production of each; printing with moveable type would never have caught on if the number of characters had been in the thousands (as in Eastern ideogrammatic languages).[11]

In the early days of computing, when the iron law of small data still ruled, the leanness of the alphabetical notation still offered considerable advantages: the first American Standard Code for Information Interchange (ASCII), back in the early 1960s, allowed for 128 slots—far more than the Latin alphabet needed, even taking into account a separate slot for capital letters, and one for each punctuation or diacritic mark; the same code was later extended to eight bits, or 256 characters, and further declined into regional variations, thus allowing for such tidbits as the euro sign (ASCII 128) or a Cross of Lorraine (ASCII 135, both in the version known as ISO Latin-1). Most keyboards today can directly generate many more signs above and beyond the double set they show, inherited from mechanical typewriters (minuscule and majuscule, or lowercase and uppercase, which originally referred to the cases where compositors stored the respective fonts).

That, however, is the past. For some years now the Gmail online interface has been supporting, alongside letters and numbers, a list of pictorial icons, known by the Japanese term *emojis* (pictograms) to distinguish them from the old emoticons (icons representing facial expressions, such as :-) or :-(, obtained through the combination of typographical signs from a standard keyboard). Unlike emoticons, emojis are ready-made pictorial images, but like letters of the alphabet in the ASCII code, each emoji is transmitted across platforms as a coded number, so each emoji number 1F638 (Unicode) will show up as something similar to a grinning cat, regardless of platform or operating system—just like the alphabet letter number 65 (ASCII) will show up as something recognizable as a capital A in all fonts, software, and graphic interfaces. In technical terms, each emoji is hence functionally similar to a letter of the alphabet, except emojis do not convey sounds but ideas (and emojical ideas are conveyed through icons, not symbols). There is, however, another purely quantitative difference: the letters of the Latin alphabet, born and bred in times of small data, were around thirty. The number of emojis in my Gmail interface, at the time of writing, is more than seven hundred and counting; my Android-based Samsung phone already offers almost 1,000 (for use in text messages as well as email); the Japanese smartphone application Line lists 46,000 (although most do not correspond to any standard code, hence they may not work outside of that company's proprietary application).

There was a time when we thought (and many feared) that electronic communications would replace print and that the digital would kill the book. The jury on that may still be out, but the frontier has already shifted: today's digital tools are phasing out alphabetical script and reviving ideogrammatic writing in its stead. For many centuries, the small-data logic of the alphabet offered its (mostly Western) users overwhelming competitive advantages: with the alphabetical notation came print with moveable type, then the typewriter. We used to think that one cannot easily type from a keyboard with thousands of keys. Today, many people—particularly young people, and not only in the Far East—are cheerfully doing just that. The alphabet is a data-compression technology that today's digital tools no longer need. With the decline of the alphabet, one of the arguments often cited by Whig historiography to account for the technical supremacy of the West in modern and contemporary history is gone too—never mind if the argument was ever sound, it simply does not apply in today's postmechanical and data-opulent environment.[12] When information technology can process almost any amount of data instantaneously and at the same cost, a notational system limited to a combination of thirty signs may seem an unreasonable constraint—a solution to a problem that no longer exists.

But the logic of Big Data will not stop there, and today's revenge of the data-opulent ideogrammatic writing against its ascetic alphabetic antagonist may in turn be short-lived. Even keyboards with thousands of virtual keys, and search tools to navigate them, may soon be replaced by speech recognition technologies. In many simple cases (searches on keywords, multiple-choice questions) the keyboard has already been replaced by voice commands, and in some cases by gesture recognition: the conversion from voice or gesture to writing is still carried out by the machine, somewhere—at least for the time being—but unbeknownst to the end-user, and without human intervention. Indeed, the keyboard (whether real or virtual, simulated by screens or tactile tools), is no longer the only human–machine interface, and soon it may no longer be the principal one. And when all script is phased out, digital tools will have gone full circle: after demolishing—in an orderly chronological regression from the more recent to the earliest—all small-data cultural technologies invented over time, humankind will be restored to something akin to the auroral primacy of gesture and the word. Digitally enhanced orality and gesture will be different from the ancestral ones, however, because voice and motion can now be recorded, notated, transmitted, processed, and searched, at will and at once; thus making all cultural technologies developed over time to handle one or the other of these specific tasks equally unnecessary.

2.2 Don't Sort: Search

On April 1, 2004, Google rolled out a free, advertising-supported e-mail service. We now know that was not one of the April Fools' Day pranks for which Google was, and still is, famous: twelve years after its launch, by invitation only, to a beta (test) version, Gmail offers fifteen GB of free storage to almost one billion users around the world. When Gmail started, however, Google was still a one-product company, known primarily for the almost miraculous efficacy of its search algorithm; so it stands to reason that Gmail was released touting full-text searchability as its main asset: Gmail's original logo, next to the beta version disclaimer, featured the now famous tagline: "**Search don't sort**. Use Google search to **find the exact message**" (emphasis in the original). Gmail remained, nominally, in beta version until 2009, which is also when the "search don't sort" motto appears to have been removed from the banner on its portal.[13] We can only assume that by that time most Gmail users had gotten used to searching instead of sorting, even though they were never really coerced into doing so; Gmail has always offered some discreet sorting tools (called "labels" rather than "folders"), and more have been introduced recently. Why, then, would Gmail users search rather than sort, and what does the philosophy of searching instead of sorting betoken, imply, and portend?

In the early days of personal computing, email programs downloaded messages from a central server to the user's machine, where messages were stored in resident folders. As often happens in the history of technical change, the new technology at first simulated the older one it was replacing, and most users simply carried over their sorting habits from their physical offices to their new virtual desktop: the interfaces of early email services reproduced a traditional filing cabinet, with folders or drawers, where users would sort and keep their documents following an order they knew, so they knew where to look for them when needed. Electronic folders could be labeled, just like paper ones, and inside each folder items were automatically sorted chronologically or alphabetically, by sender or by the title in the subject line. But early personal computers offered limited searching tools, if any, so finding a document after putting it in a folder was a time-consuming and hit-or-miss affair, much dependent on memory or other idiosyncratic factors. When the number of email messages started to surge, well beyond the traditional limits of scribal or even mechanical letter writing, manual sorting struggled to cope. Then came Gmail.

Gmail made the full text of all email messages, sent or received, searchable for words and numbers using a simplified Google syntax: to this day, search results are not ranked by relevance, but can be filtered by domain or chronology. Alphanumerical Gmail searches can only be performed online, hence users are encouraged to use Gmail primarily as a web-based service. At the beginning Google also prompted users never to delete messages, and to this day it is not known precisely what happens to messages after they are permanently deleted, or how long they survive on the company's proprietary backup systems—an open-endedness that many find philosophically and legally disturbing.[14] Anything not permanently deleted, however, is searchable, and the "search, don't sort" logic suggests that an automated full-text search for words or numbers on a whole corpus of sources, left raw and unsorted, is a more powerful retrieval tool than the traditional manual process of first sorting items by topic and then looking for them in their respective folders.

I cannot find any scientific enquiry to validate that assumption, but plenty of anecdotal evidence does. "Search don't sort" certainly works for me. Yielding to the new technical logic of the tool, I diligently phased out most folders over time (while still using a handful of Gmail "labels," occasionally and for specific projects). Now, when I need to find a message, I do not look for it in a given place (or folder), but search for a combination of words, numbers, dates, and people, which I remember as more or less related to the matter at hand. While I wish at times that the search would offer a more advanced syntax, most of the time it works—and it certainly works better than my own sorting would, even if I had adopted a perfect taxonomy from the start, and unremittingly and consistently complied with it for the last ten years. Thus, over time, I have become well advanced in the new digital art of finding stuff without knowing where it is.

From the beginning of time, humankind has conceived and honed classifications for two main reasons: as a way to find or make some order in the world—an idea often cherished by philosophers, theologians, and thinkers and which some see as a universal human yearning—and, more simply, to assuage the basic human need to put things in certain places, so we know where they are when we need them. We used to think that sorting saves time. It did; but it doesn't any more, because Google searches (in this instance, Gmail searches) now work faster and better. So taxonomies, at least in their more practical, utilitarian mode—as an information retrieval tool—are now useless. And of course computers do not have queries on the meaning of life, so they do not need taxonomies to make sense of the world, either—as we do, or did.

Taxonomies are as old as Western philosophy: one of the most influential dates back to Aristotle, who divided all names (or all things, the jury is still out) into ten categories, known in the Middle Ages as *predicamenta*.[15] Aristotle's classifications do not seem to have directly influenced the first encyclopedists (Varro, Pliny, or Isidorus), but with Medieval Scholasticism the division of names into classes merged with another Aristotelian tenet, the diairetic structure of universals and singulars, genus and species, which today we more often interpret in terms of set theory: classes are constituted by all items that have something in common, with smaller sets sharing more predicates, all the way to the individual, or set of one (which is where, in modern terms, maximum definition meets minimum extension).[16] With classifications construed as cascades of divisions and subdivisions, encyclopedias (which originally, and etymologically, referred to a circle, or cycle, of knowledge) started to look less like circles and more like trees, with nested subdivisions represented as branches and sub-branches.

The first "tree of knowledge" is attributed to the Majorcan philosopher and Catalan writer Ramon Llull (Raymond Lull, or Lullus, late thirteenth century), also known as a precursor of Leibniz's *ars combinatoria*.[17] Branching taxonomies flourished in the sixteenth century, both as a philosophical instrument and as an indexing tool for librarians: the French logician and pedagogist Pierre de la Ramée (Petrus Ramus, which, incidentally, means "branch" in Latin) thought that every argument, subject, or idea could be visualized as an arborescent diagram of definitions and subdivisions, and he started to implement a universal program of pedagogical reform where every discipline was represented as a tree, endlessly divided into branches. Ramus also thought that the same treelike structure should apply to every discourse on every subject, in poetry and in prose, in writing as well as in speech. This may seem a bit extreme today,[18] and Ramus's "universal method" (*Methodus Unica*) may have proven less than popular among some students and staff back then: Ramus himself was stabbed and beheaded during the first night of the Saint Bartholomew's massacre, his body thrown out of the window of his room at the Collège de Presles and dumped into the Seine. Almost at the same time, very similar treelike diagrams were used by the Swiss physician, botanist, and bibliographer Conrad Gessner, who published several versions of a general catalog of all printed books he knew (and some manuscripts to boot), some ordered alphabetically and some by subject.[19] By the mid-sixteenth century, with book production surging due to print, librarians and book dealers might have been facing the first big data crisis in history, as—much like today—many of their traditional tools and practices were being overwhelmed by technological change. To help keep track of this unprecedented wave of authors and titles, "instantly taken up, multiplied and spread far and wide by printing as by a superhuman war machine,"[20] Gessner devised a universal method of classification by treelike subdivisions pegged to a numeric ordering system—an index key which could also have served as a call number, although it does not seem it ever was. That could hardly have happened in the south of Europe anyway, as all of Gessner's books, particularly his bibliographies, were immediately put in the Index of Prohibited Books by the Roman Inquisition.[21]

One generation before, another early visionary of the mechanization of knowledge found other ways to cope with the growing inadequacy of the information retrieval tools of his day. Giulio Camillo's famous "Memory Theatre" was a filing cabinet where all extant writings by Cicero were sorted by subject and further subdivided by size; Camillo's idea was that modern writers could compose all sorts of new texts—indeed, all possible texts—by cutting and pasting suitable snippets excerpted from this systematic archive of Cicero's perfect language and universal wisdom. Camillo designed some ingenious data retrieval tools for that purpose, but as none of these really worked he resorted to occultism and divination, suggesting that his Ciceronian machine would only work in full for the initiated. Camillo was also known during his lifetime as an accomplished magician and a hypnotizer of lions.[22]

Thus it will be seen that our dominant sorting tool, the alphabetical classification, was somewhat of a late bloomer: early dictionaries in print were often ordered thematically, not alphabetically (i.e., they were in fact thesauri), and Ramus's and Gessner's arborescent classifications survive in all the bibliographic tools still in use today, such as the Universal Decimal Classification (UDC), Library of Congress, or Dewey classification systems. Paul Otlet's UDC was also the springboard for the project of the Répertoire Bibliographique Universel—an open-ended index of everything, which Otlet kept expanding until 1934, when it reached fifteen million index cards, and never served any practical purpose.[23] A few years ago Google announced a partnership with the Mundaneum, the museum in Mons, Belgium, dedicated to the legacy of Otlet's life project,[24] and there may be some logic in the fact that Google, which has already changed the world with digital search engines, should celebrate the memory of the man who brought mechanical sorting technologies to their all-time, perfectly useless zenith.

In 1530, when Camillo needed a suitable taxonomy to sort his archive of Cicero's words, phrases, and arguments, he looked high and low for a classification scheme that would be, in his words, "discrete and capacious as needed, and furthermore able to keep the mind awakened and to imprint itself into memory."[25] Taxonomies are made by and for the human the mind, so they should be intelligible and memorable. From today's big data perspective, however, it is easy to see that classifications also function by structuring and formalizing a random stock of information, providing speedier access to all data in an inventory by way of index keys. Thus, by the logic it provides, an arborescent structure is easier to manage, use, and memorize than the raw aggregate of data it refers to; but the same structure also functions as an orientation tool that allows end-users to calculate the coordinates of each item in stock. So, for example, in the Library of Congress classification, N is the general call number for all the fine arts, NA for architecture, NA 190–1555 for architectural history, and NA 510–589 for the history of architecture in the Renaissance (etc.); if we know how the code works, we can walk straight to the shelf where the subject we are looking for is kept, without having to read the titles of all the books on all shelves. And, even if we do not often think of it that way, the coordinates of each word in a dictionary are calculated by running each letter in that word against a very simple code—the alphabetical sequence. Thus the word "abacus," for example, has coordinates 1,2,1,3,21,19; and those coordinates, applied sequentially, lead us straight to that word in the dictionary without having to read any other. Thanks to alphabetic sorting, instead of remembering the order of every word in a dictionary, we only need remember the order of the alphabet itself.

But computers do not work that way. To search for the word "abacus" in a corpus of textual data, computers will scan the whole corpus looking for a given sequence of forty-eight 0s and 1s, and stop whenever that sequence shows up—regardless of how that corpus may or may not have been sorted. Computers search, they don't sort. More and more often, so do we.[26] Gmail is training us to leave our documents unsorted because computers search faster than we sort, but this principle need not be limited to media objects; it can be extended to physical objects

65

P. RAMI DIALECTICA.

TABVLA GENERALIS.

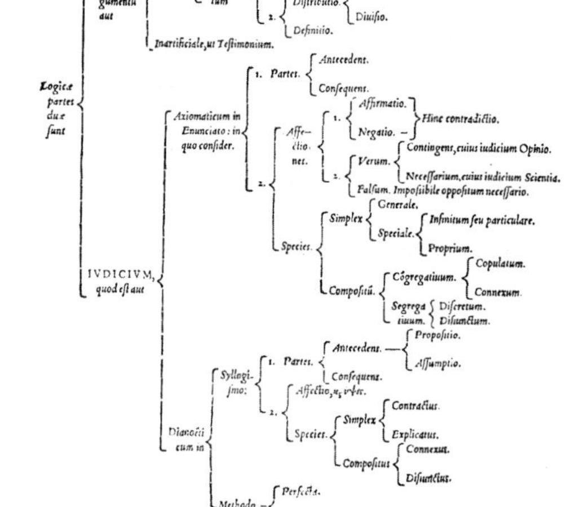

2.2

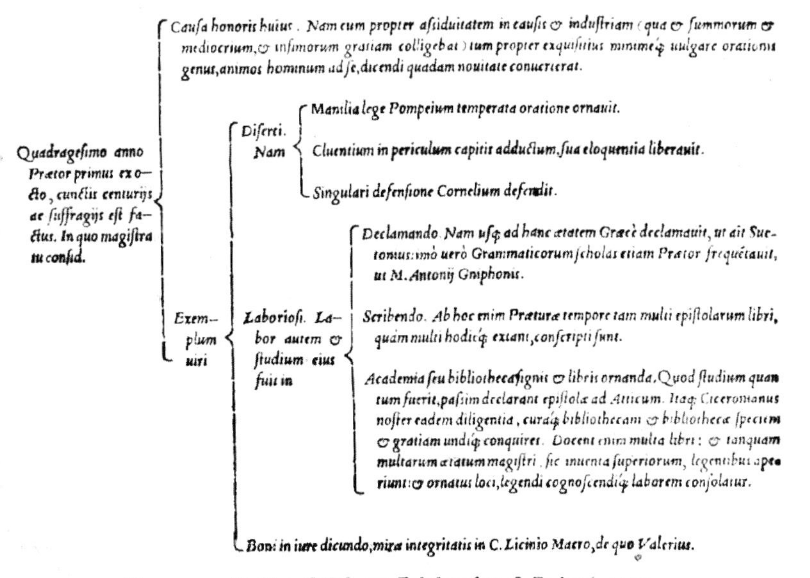

2.3

Arborescent diagrams (figures 2.3 and 2.4) showing the classification of the art of logic in Pierre de la Ramée's (Ramus's) *Dialectica*, from Ramus's *Professio Regia*, published and edited posthumously by Johann Thomas Freige, where similar diagrams are also used to classify the history and timeline of Cicero's main life events (here, 2.3, outlining Cicero's prateorian election at the age of 40). Petrus Ramus (Pierre de la Ramée), *Professio Regia, Hoc est Septem artes liberales, in Regia Cathedra, per ipsum Parisiis apodictico docendi genere propositae, et per Ioan. Thomam Freigium in tabulas perpectuas ... relatae* (Basle: Sebastian Henricpetri, 1576).

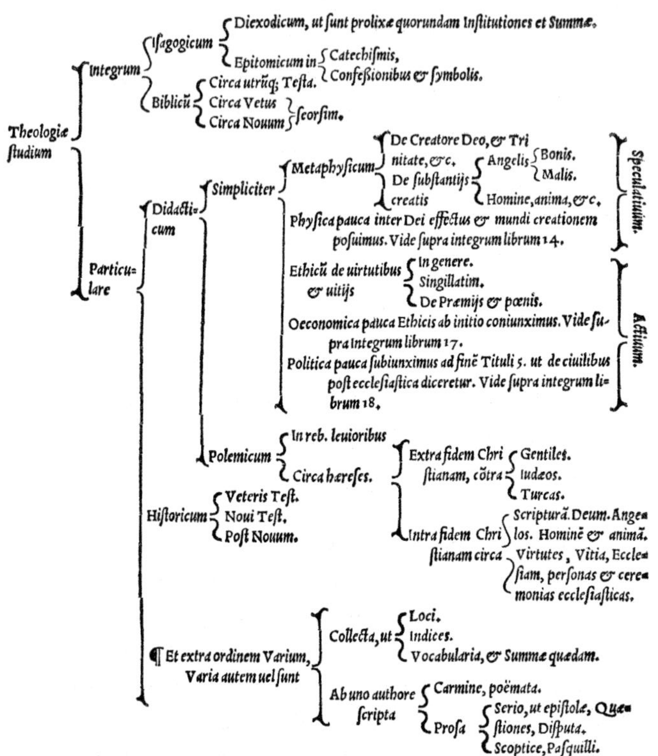

2.4

Conrad (Konrad) Gessner (Gesner), diagram showing part of the classification of the science of theology, from *Partitiones theologicae: pandectarum universalium Conradi Gesneri liber ultimus …*, (Zurich: Froschauer, 1549). © Zentralbibliothek Zürich, (Shelf mark: 5.13, 3).

of all kinds, which can be tagged and tracked using Radio Frequency Identification (RFID). This may apply to random junk in a garage, to books in a library, or to the full inventory of Amazon.com. Indeed, all kinds of items in Amazon warehouses, including books, are not sorted by subject or category, but only based on the frequency of sale, following an order that would be meaningless to a human mind.[27] Using the same technical logic, in our houses we could keep potatoes and socks and books and fish in the same drawers, or indeed anywhere. We would not need to remember where we put things, or where things are, because whenever we need them we could simply Google them in physical space—and see the result on a screen, on a wristwatch, or through augmented reality eyewear. Let's go one step further: the same Google logic need not be limited to space—it may also apply to time.

2.3 The End of Modern Science

Many sciences, philosophies, and systems of thought look at the past to try and predict the future, because humans themselves learn from experience—and according to some, from experience only. Be that as it may, modern experimental science is based on the assumption that events that tend to repeat themselves with a certain regularity may be predicted. For example, if every time we drop a stone it falls to the ground, we expect that the next time we drop a similar stone in similar conditions it will fall in a similar way, and from the regularities we observe in the repetition of this process we extrapolate rules that will help us predict other comparable incidents before they happen. This is the way modern science (inductive, inferential, experimental science) has worked so far, with great success; the rules this science arrives at are typically expressed as mathematical formulas.

Yet, once again, our data-rich present prompts us to look at our data-starved past from a different vantage point. In terms

2.5
Andreas Gursky, *Amazon* (2016). © Andreas Gursky.
Courtesy: Sprüth Magers Berlin London / DACS 2016.

of pure quantitative data analysis, the history of the modern scientific method (and, indeed, of modern science as a whole) appears as little more than a succession of computational strategies strenuously developed over time to maximize the amount of information we could collect, record, and transmit for scientific (i.e., predictive) purposes. Evidently, the best way to learn from experience would be to keep track of each experiment in its totality. As nobody knows how to define the totality of an event, limits must always be set so the collection of data can stop at some point. However, given the drastic limitations in the amount of data we could afford to keep and process in the past, scientists had to be picky, and turning necessity into virtue, they learned to focus on only a few privileged features for each experiment, discarding all others. Thus over time we learned to extrapolate, generalize, infer, or induct formal patterns, and we began to record and transmit condensed and simplified interpretations of the facts we observed, instead of longer, fuller descriptions of the facts themselves. Theories tend to be shorter than the description of the events they refer to,[28] and indeed syllogisms, then equations, then mathematical functions, were, and still are, very effective technologies for data compression. They compress a long inventory of events that happened in the past into a very short script, generally in the format of a causal relationship, from which we can deduct many more events of the same kind, if similarly structured—past or future indifferently.

Galileo, one the founding fathers of modern science, also laid the foundations of modern mechanics, static, and of modern structural engineering. In his last book, *Two New Sciences*, which had to be smuggled out of Tuscany and printed in the Dutch Republic to escape censorship, Galileo famously described a number of experiments he had made to study how beams break under load.[29] But today we need not repeat any of his experiments, or any other experiment, to determine how standard beams will

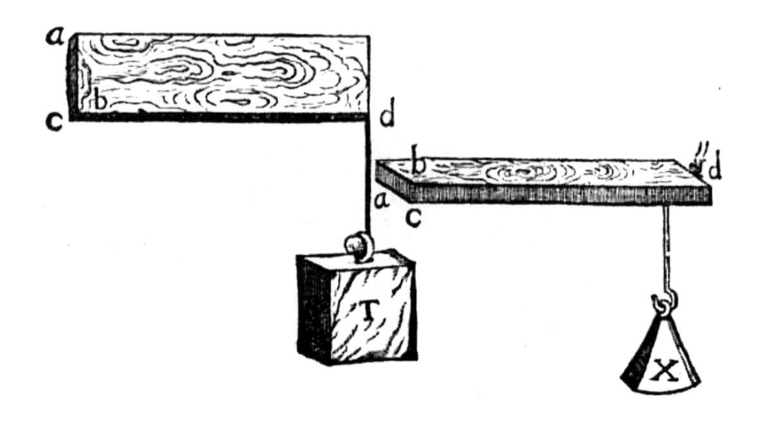

2.6

Galileo Galilei, experiments prefiguring Leonhard Euler's (1707–1783) moment of inertia, (2.6), and on the deflexion of a beam supported on both ends, and cantilevered (2.7), from *Discorsi e Dimostrazioni Matematiche intorno à due nuove scienze, attinenti alla mecanica e i movimenti locali* (Leiden: Elzevir, 1638).

quenza l'altra E F *fiſſo il termine* F, *è manifeſto, che poſti i ſoſtegni*

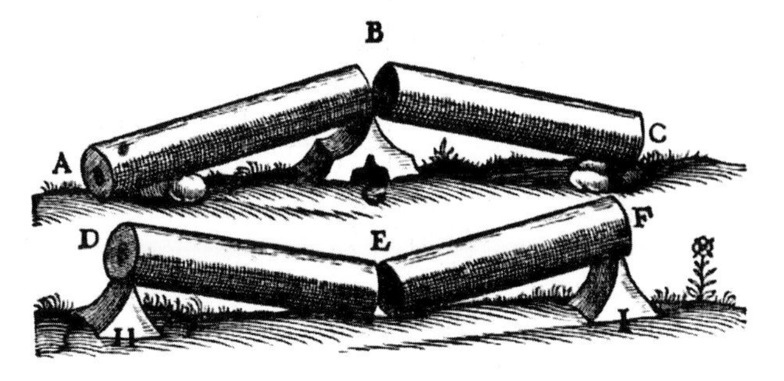

H I *ſotto l'eſtremità* D F, *ogni momento che ſi aggiunga di forza, ò di peſo in* E, *quiui ſi farà la rottura.*

2.7

break under most standard loads because, generalizing from Galileo's experiments and from many more that followed, we have obtained a handful of laws that all engineers study in school: a few, clean lines of mathematical script, easy to grasp and commit to memory. These formulas, or rules, derive from regularities detected in many actual beams that broke in the past and predict how most beams will break in similar conditions in the future.

However, let's imagine, again, that we live in an ideal big data world, where we can collect an almost infinite amount of data, keep it forever, and search it at will, at almost no cost. Given such unlikely conditions, we could assume that every one of those experiments, or more generally, the experiential breaking of every beam that ever broke, could be notated, measured, and recorded. In that case, for most future events we are trying to predict, we could expect to find and retrieve a precedent, and the account of that past event would allow us to predict a forthcoming one—without any mathematical formula, function, or calculation. The spirit of the new science of searching, if there is one, is probably quite a simple one, and it reads like this: *whatever happened before, if it has been recorded, and if it can be retrieved, will simply happen again, whenever the same conditions reoccur*. This is not different from what Galileo and Newton thought. But Galileo and Newton did not have Big Data; in fact, they often had to work from very few data indeed. Today, to the contrary, instead of calculating predictions based on mathematical laws and formulas, the way they did, we can simply search for a precedent for the case we are trying to predict, retrieve it from the almost infinite, universal archive of all relevant precedents that ever took place, and replicate it.[30] When this happens, search can replace the method of modern science in its entirety. This may sound like a joke—or a colossal exaggeration. Quite to the contrary, in pure epistemological terms, the new science of search is little more than a marginal quantitative upgrade in the conceptual tools at our disposal.

Albeit ostensibly predictive, modern science does not really predict the future—it never did. The scientific formulas we use to deduct and calculate events before they happen are inferred and extrapolated from a number of experiments already made, and the only thing those formulas can really do is to refer back to, and in fact retrieve, the experiments from which they derive. We may well imagine that some of those events, or some patterns derived from them, will repeat themselves in the future, because they have repeated themselves so many times in the past; but that we do by a leap of faith: the only certainty we have is in the evidence that has already been served. Thus science is always a retrieval, never a prediction. The old science of small data retrieved events that had been recorded in an abbreviated, condensed, compressed, or truncated form due to the limits in the old technology of data recording and transmission. Today's new science of Big Data has removed many of those limits, so instead of retrieving pale shadows we can now retrieve vivid 3-D avatars of the original facts (or something as close to that as the amount of available data allows). While the old science of small data used scientific formulas to deduct (but in fact to retrieve) a skinny, pitiful handful of numbers, the new science of Big Data can search and retrieve full-bodied, hi-res, high-definition precedents almost in their entirety.[31] Either way, we bring evidence back from the past in the hope this may help us predict future events; but while in the old days we used the human science of sorting for that purpose, today we can use the computational art of searching instead. Furthermore, in today's computational environment, the precedent we dig up need not be a real one; it may equally have been simulated on purpose. This is what avant-garde designers have been doing for the last few years, thanks to the power of today's digital computation—only in some slightly different ways, and sometimes without saying it in so many words.

2.4 The New Science of Form-Searching

Recent technical developments in the extrusion and robotic winding and weaving of very thin filaments have prompted exciting and promising experiments—at the Institute for Computational Design (ICD) of the University of Stuttgart, at the Bartlett (University College London), and elsewhere.[32] In 2014, Achim Menges and Jan Knippers published a groundbreaking technical article describing their use of fiber-reinforced polymers in the thin shell of the experimental ICD/ITKE Research Pavilion they built in 2012.[33] Structural calculations for the pavilion had to take into account the complex geometry of the shell, as well as the density and direction of each bundle and layer of carbon and glass fibers wound in it. The authors began with a geometrical and material layout inspired by biological models; then they simulated the structural behavior of this first model using standard computational finite element analysis (FEA), a mathematical method for the calculation of deformation and stresses within a continuous structure.[34] Based on the results of this first simulation, some aspects of the design were tentatively tweaked, altering both the geometry of the shell and the internal layout of the fibers. The FEA simulation was then rerun on this second model, and so on, and repeated (iterated) many times over until the authors were pleased with the results.

In this process of heuristic (not mathematical) optimization,[35] every simulated model that was tried and discarded corresponded to a physical model that a traditional artisan would have made, tested, and likely broken in real life. Using digital simulations of structural performance, however, today we can make and break on the screen in a few hours more full-size trials than a traditional craftsman would have made and broken in a lifetime.

2.8
Alisa Andrasek, Wonderlab, AD Research Cluster 1, B-Pro M.Arch Architectural Design, The Bartlett UCL, *Liquid* (2016). Robotic extrusion of filaments on a vectorial template derived from computational fluid dynamics. Tutors: Alisa Andrasek, Daghan Cam, Andy Lomas. Robotics: Feng Zhou. Students: Zhuoxing Gu, Tianyuan Xie, Bingyang Su, Anqi Zheng. © Alisa Andrasek, AD Research Cluster 1, The Bartlett UCL.

2.9

Gilles Retsin, Manuel Jimenez-Garcia, AD Research Cluster 4, B-Pro M.Arch Architectural Design, The Bartlett UCL, *CurVoxels, 3D Printed Chair* (2015). A robotically extruded chair combining a curved toolpath with a voxel-based data structure. Students: Hyunchul Kwon, Amreen Kaleel, Xiaolin Yi.

2.10
Gilles Retsin and SoftKill Design, *Protohouse*, Collection Centre Pompidou (2012). Structural optimization (by iterative removal and addition of material) aimed at obtaining minimal volume and uniform stress throughout a complex architectural envelope. SoftKill design team: Nicholette Chan, Gilles Retsin, Sophia Tang, Aaron Silver. Developed at the Architectural Association Design Research Lab in London.

2.11
ICD Institute for Computational Design (Prof. Achim Menges). ITKE Institute of Building Structures and Structural Design (Prof. Jan Knippers), *ICD/ITKE Research Pavilion 2012*, Robotic filament winding of carbon/glass fiber structure, University of Stuttgart, 2012. © ICD/ITKE University of Stuttgart.

2.12
ICD Institute for Computational Design (Prof. Achim Menges). ITKE Institute of Building Structures and Structural Design (Prof. Jan Knippers). *ICD/ITKE Research Pavilion 2012,* Exterior view of pavilion, University of Stuttgart, 2012. © ICD/ITKE University of Stuttgart.

Artisans of pre-industrial times (as well as the ideal artisan of all times recently romanticized by Richard Sennett)[36] were not engineers; hence they did not use mathematics to predict the behavior of the structures they made. When they had talent they learned intuitively, by trial and error, by making and breaking as many samples as possible. So do we today, using iterative digital simulations. We may or may not intuit some pattern, regularity, or logic inherent or embedded in the structure we are tweaking—but that is irrelevant. By making and breaking (in simulation) a huge number of variations, at some point we shall find one that does not break, and that will be the good one.

Inspired by Frei Otto's method of physical form-finding—which Menges was the first to implement digitally and to translate into computational terms—this heuristic design process is functionally equivalent to the big data, search-based alternative to modern science mentioned earlier. Whenever digital tools allow us to collect, record, and process huge troves of data, information retrieval (the search for a precedent) is more effective than the traditional, deductive application of scientific formulas or any other law of causation. The 2012 ICD/ITKE Research Pavilion being an experiment without precedent, no corpus of previous, comparable instances was available for search and retrieval. In the absence of any such historical archive, however, Menges, Knippers, and their team could avail themselves of the immense power of digital simulation to create on the spot, virtually, just the archive they would have needed. They may not have seen it this way, but by simulation and iteration they generated a vast and partly random corpus of many very similar structures that all failed under certain conditions; and they chose and ultimately replicated one that did not.[37] This is a far cry from how a modern engineer would have designed that structure—which is one reason why no modern engineer could have designed it.

A modern engineer would have started with a set of formulas establishing causal relationships between loads, forms, and stresses in the structure. Typically used to calculate the resistance of a structure after it has been designed, these formulas can also drive and inspire our first, intuitive design of it. This is because causal laws make sense, somehow: by the causality they express, they interpret and provide some understanding of the physical phenomena they describe. Indeed, in the classical scheme of things, causality is seen as a primary law of nature; so the laws of mechanics, for example, are held to spell out in mathematical terms the way beams, cantilevers, pillars, arches, or vaults function in reality, and the formulas of structural engineering have a "meaning" which is held to be true to nature. Indeed, this meaningfulness, and the structural theories from which it derives, are visible in all masterpieces of modern structural engineering, from Eiffel's tower to Nervi's vaults. If we look at these modern structures we understand the basic structural principles their designers had in mind when they first sketched them.

That does not apply to our current way of designing by form-finding, or, as we should perhaps say, to better demarcate the nature of today's process from that of its physical precursor, computational form-searching. The power of Big Data applied to information retrieval, simulation, and optimization makes the formulaic data compression at the core of modern structural engineering as obsolete as the Yellow Pages—or as the logarithmic tables mentioned above. Gilles Deleuze famously disparaged the abstract determinism of modern science, to which he opposed the heuristic lore of artisan "nomad sciences."[38] Once again, Deleuze's view of our pre-mechanical past doubles as an eerily cogent anticipation of our digital future. Through computational form-searching we can already design new structures of unimaginable complexity. But precisely because it is unimaginable, this posthuman complexity belies interpretation and transcends the small-data logic of causality and determinism we have invented over time to simplify nature and convert it into reassuring, transparent, human-friendly causal models. Why does one so unimaginably complex structure work, and the thousands of very similar ones we just ran through FEA simulation do not? Who knows. But the point is that it works. And if that is the case, then we must come to the almost inevitable conclusion that the new science of search may soon replace the method of experimental science in its entirety, simply because simulation and search can solve problems that the formalistic approach of modern science could never tackle. Computers can search faster than humans can sort.

Digitally intelligent designers may be more enthusiastic or more outspoken than other early adopters, but the new science of search has already pervaded, in spirit if not in letter, many of today's data-driven cultural technologies, and traces of the same quantitative, heuristic use of data are evident, in some muted, embryonic way, in other branches of the natural sciences, such as weather forecasting.[39] And sure enough, some historians of science have already started to investigate the matter—with much perplexity and reservation, as one would expect; as the postmodern science of big data and computation marks a major shift in the history of the scientific method.[40] As mentioned above, mathematical abstractions such as the laws of mechanics or of gravitation, for example, or any other grand theory of causation, are not only practical tools of prediction, but also, and perhaps first and foremost, ways for the human mind to make sense of the world. But then, if abstraction and formalization (that is, most of classical and modern science, in the Aristotelian and Galilean tradition) are also seen as contingent and time-specific data-compression technologies, one could argue that in the absence of the technical need to compress data in that particular way, the human mind can find many other ways to relate to, or interpret, nature. Epics, myth, religion, and magic offer vivid historical examples of alternative, nonscientific methods, and no one can prove that the human mind is, or ever was, hard-wired for modern experimental science. Many postmodern philosophers, for example, would strongly object to that notion. And as so many alternatives to modern science existed in the past, one could argue that plenty of new ones may be equally possible in the future.

The mere technical logic of the new science of searching goes counter to core postulates and assumptions of modern science. Additional evidence of an even deeper rift between the two methods is easy to gather. Western science used to apply causality to bigger and bigger groups, or sets, or classes of events—and the bigger the group, the more powerful, the more elegant, the more universal the laws that applied to it. Science, as we knew it, tended to universal laws—laws that bear on as many different cases as possible. The new science of data is just the opposite: using information retrieval and the search for precedent, data-driven prediction works best when the sets it refers to are the smallest. Indeed, searches are most effective when they can target and retrieve a specific and individual case—the one we are looking for.[41] In that, too, the new science of data represents a complete reversal of the classical (Aristotelian, Scholastic, and early modern) scientific tradition, which held that individual events cannot be the object of science.[42]

In social science and in economics, this novel data-driven granularity means that instead of referring to generic groups, social and economic metrics can and will increasingly relate to individual cases. This presages a brave new world where standards and averages will no longer be either required or warranted: fixed prices, for example, which were introduced during the Industrial Revolution in order to standardize retail transactions, have already ceased to exist, as each retail transaction in a digital marketplace today is an algorithmically customized one-off, delivered at zero processing costs, or almost.[43] Likewise, the cost of medical insurance, calculated as it still is on the basis of actuarial and statistical averages, could become irrelevant, because it may be possible to predict, at the granular level, that some individuals will never have medical expenses, hence they will never need any medical insurance, and some will have too many medical expenses, hence no one will ever sell them any medical insurance. The individual that is the object of this new science of granular prediction will no longer be a statistical abstraction—it will become each of us, individually. This may be problematic from a philosophical and religious point of view, as it challenges traditional ideas of determinism and free will; but in more practical terms, it is also incompatible with most principles of a liberal society and of a market economy in the traditional, modern sense of both terms.

Natural sciences, however, offer quite a different picture. As recent works by Neri Oxman and others have shown, we can now design and fabricate materials with variable properties at minuscule, almost molecular scales; and we can detect, quantify, and take into account the infinite, minute, and accidental variations embedded in all natural materials—a capriciousness that made natural materials unsuitable for industrial use, and almost eliminated them from modern industrial design.[44] Artisans of pre-industrial times did not have much choice: they had to make do with whatever natural materials they could find. For example, when Alpine farmers had to roof a new chalet, they looked high and low (literally) for a tree that would be a good fit for the ridge piece; sometimes the shape of the roof would be tweaked to match the quirks of the best available trunk. And cabinetmakers could (and the extant few still can) skillfully work around almost any irregularity they find in a plank of timber and make the most out of it. But industrial mass production follows a different logic. To be used in an assembly line, or pieced together by unskilled workers, timber must be processed and converted into a homogeneous material compliant with industrial standards—as plywood, for example, which is a factory-made industrial product, although derived from wood. Artisan masons of old (and few survive in the so-called industrialized countries) knew very well how to make concrete on site the smart way, making it stronger, for example, in the angles and corner walls (more cement), cheaper in some infill (more rubble), thinner and more polished next to some openings (more sand), etc. But for engineers, concrete had to be dumb, homogeneous, and standard, the same all over, because that was the only material they could design with the notational and mathematical tools at their disposal. Even assuming an engineer could calculate the structural behavior of variable property concrete (concrete with different performances in different parts of the same structure), until recently there was no practical way to produce those variations to specifications, either by hand or mechanically. After all, reinforced concrete is only an elementary, two-property material, yet it took several decades to learn a consistent, reliable way to design, calculate, fabricate, and deliver it.

In theory, and increasingly in practice, digital design and fabrication tools are eliminating many of the constraints that came with the rise of industrial standards. X-ray log scanning, for example, is already used in forestry: trees are scanned prior to felling, and the cutting of the boards is customized for each trunk to minimize waste. The scan is discarded by the sawmill after the planks are sold, but there is no reason not to envisage a full design-to-delivery workflow, in this case extended to include the natural production of the source material—from the forest to the end-product, perhaps from the day the tree is planted (which would once again curiously emulate ancestral practices of our pre-industrial past).[45] Each tree could then be felled for a specific task: a perfect one-to-one match of supply and demand that would generate economies without the need for scale—which is what digital technologies typically do when they are used the right way. Likewise, variable property materials can now be designed and fabricated at previously unimaginable levels or resolution, including concrete, which can be extruded and laid by nozzles on robotic arms, so each volumetric unit of material can be made different from all others. This is what artisanal concrete always was—which always scared engineers to death, because they could not design and calculate that. Today we can.

Much as modern science tended to more and more general laws bearing on the largest possible sets of events, modern technology tended to the mass production of standardized materials that were designed to be, as much as possible, homogenous and isotropic. Industrial standards were meant to generate economies of scale, but also, and crucially, homogenous materials could be described and modeled using elegant mathematical tools such as differential equations and calculus.[46] Calculus is a mathematics of continuity, which abhors singularities: it is perfect, for example, to quantify the elastic deformation of any homogeneous chunk of continuous matter. That is why the modern science of engineering can calculate the stress and deformations of the Eiffel Tower, which is made of iron, but until recently the same science could not calculate the resistance of a ten-foot-high brick-and-mortar garden wall.

To the contrary, using digital simulation and data-driven form-searching, we can now model the structural behavior of each individual part in a hypercomplex, irregular, and discontinuous 3-D mesh. And using digital tools, we can fabricate any heteroclite mess precisely to specs, on time and on

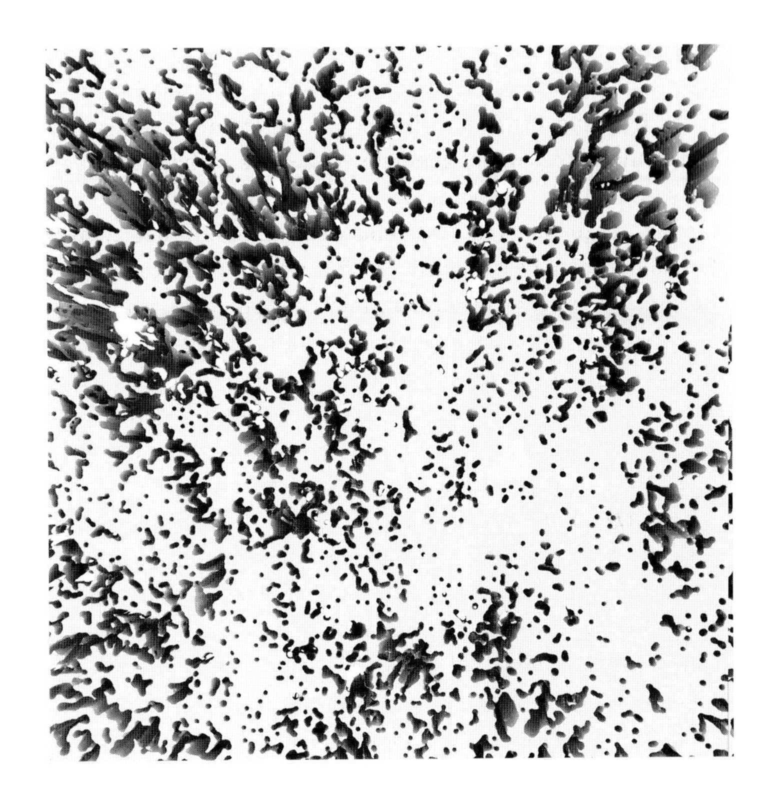

2.13
Alisa Andrasek, Wonderlab, AD Research Cluster 1, B-Pro M.Arch Architectural Design, The Bartlett UCL, *Morphocyte* (2016). Variable property materials designed and fabricated through the simulation of the biological process of cellular division. Tutors: Alisa Andrasek, Daghan Cam, Andy Lomas. Robotics: Feng Zhou. Projects/Students: Zuardin Akbar, Yuwei Jing, Ayham Kabbani, Leonidas Leonidou. © Alisa Andrasek, AD Research Cluster 1, The Bartlett UCL.

budget: robots will see to that. Industrial materials were standardized so they could be calculated and mass-produced. Today we can calculate and fabricate variations at all scales, and compose with unlimited variations as needed or as found in nature. Used this way, the new science of granular prediction does not constrain but liberates, and almost animates, inorganic matter.[47] And far from being a mere, albeit powerful, inspirational metaphor—which it has been since the start of the digital turn in architecture—vitalism is already, in many cases, an actual and perfectly functional strategy, underlying or already embedded in many experiments, tendencies, and trends that populate today's computational design.

Indeed, alongside the traditional, positivistic approach to the digital design and fabrication of variable property materials—which would push the resolution of predictive models and design notations to the highest level of granular detail compatible with the task, materials, and technology at hand—another, quite different option appears to be increasingly viable. As Menges has shown, in some cases the easiest way to cope with unwieldy or quirky materials is to devolve some capacity for adaptive improvisation to the last stage of robotic manufacturing, above and beyond traditional margins of tolerance.[48] Given the ease and speed of data collection by ubiquitous sensors during all phases of production, robotic manufacturing can already include some reactive, autonomous skills. Regardless of all practical considerations, this approach also reflects a certain idea of the physical world and of the nature of matter: if inorganic matter is alive (as some believe it is, regardless of etymology), then its behavior is also to some extent unpredictable or indeterminable, and the only way to deal with the inherent capriciousness of such "living" materials is to react to their whims and volitions on the spot and on the fly. This would once again vindicate the well-known, pervasive analogy between computational fabrication and the "smooth" tooling of traditional craftsmanship: no artisan would X-ray a piece of timber before working on it, but all good artisans would know how to make the best of whatever they find in it when they start carving it. Likewise, many expert dentists would refrain from advanced digital scanning of a tooth they must treat, and would rather keep drilling it, tentatively, until their hapless patient screams. A first and obvious technological upgrade of this truculently heuristic method would be to have the dentist 3-D scan the tooth to the highest possible resolution, then calculate the best path for the drill on the model, before surgery starts—thus turning the dentist into an engineer, and in fact into a designer, as the whole surgery would be designed in full and in advance on a digital model of the operating theater. This would be the approach of modern structural design, and of design in general, as it has been known since the Renaissance: design is a predictive tool; it models something before it happens. Yet while many patients today would undoubtedly like their dentists to behave like designers, many avant-garde designers seem to prefer their robots to behave like old-school dentists—stopping when the material screams, so to speak. And, oddly, this trial-and-error approach to adaptive and reactive robotic fabrication is already yielding more promising and more practical industrial applications than the traditional scan-and-design approach. Whether we like it or not, the future of robotics may be closer to popular quackery than to industrial engineering.

2.5 Spline Making, or the Conquest of Free Form

All tools modify the gestures of their users, and in the design professions this feedback often leaves a visible trace: when these traces become consistent and pervasive across objects, technologies, cultures, people, and places, they coalesce into the style of an age and express the spirit of a time. The second

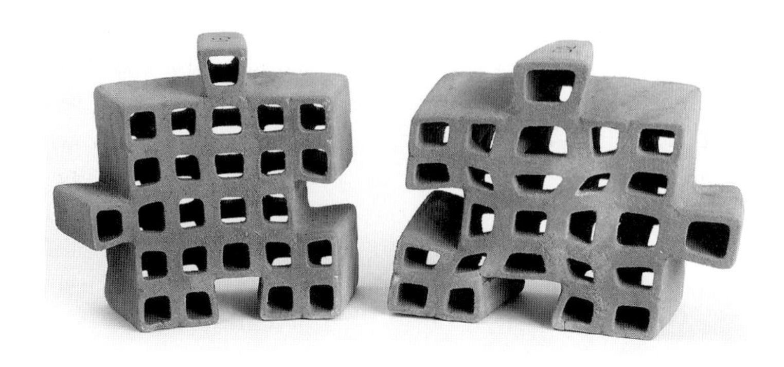

2.14

Jenny E. Sabin, *PolyBrick* (2015–16). Polybrick 1—Unfired PolyBricks featuring 3-D printed high-fire clay body. Principal investigator: Jenny E. Sabin. Design research team: Martin Miller, Nicholas Cassab, Jingyang Liu Leo, David Rosenwasser. ©Sabin Design Lab, Cornell University. Image courtesy: Cooper-Hewitt Design Museum.

digital style, the style of a data-affluent society and of a nouveau data-rich technology, is the style of the late 2010s. And, as often happens in the history of technology, a good way to assess what is distinctive in the things we make, and in the way they look, is to look at the tools we have stopped using and take stock of the things we have just stopped making. Nothing shows the small-data logic of the first digital age in architecture better than the history of its most distinctive and recognizable trait, the spline-based curve.

As we now know, the first digital style in the 1990s turned out to be one of curves—or, as designers like to say, of "spliny" curves, in reference to the mathematics of continuous curves, or splines—which was one of the novelties of early CAD/CAM, computer graphics, and animation software. Yet one would be hard pressed to find any overarching or long-lasting reason to explain why computers should be primarily—or indeed, almost exclusively—used to make sinuous lines and curving surfaces, as they were in the 1990s. In fact, the theory of digital mass customization, as it was known back then, would suggest just the opposite.[49] Starting from the early 1990s, the pioneers of digitally intelligent architecture argued that variability is the main distinctive feature of all things digital: within given technical limits, digitally mass-customized objects, all individually different, should cost no more than standardized and mass-produced ones, all identical. As computers and robots do not articulate aesthetic preferences, using CAD/CAM technologies we should be able to design and make boxes as well as blobs, as need be, at the same unit cost. Digital curvilinearity began to emerge as a theoretical trope in the mid to late 1990s, when it was seen as a side effect of sorts of the so-called Deleuze connection in architecture, in particular through the influence of Deleuze's book, *The Fold: Leibniz and the Baroque*.[50] But the theory of the *objectile* (better known today as digital parametricism), as outlined by Deleuze and Bernard Cache in that book, spoke for digital variability as a general tenet of nonstandard mass production, unrelated to any specific visual form. Deleuze's "fold" itself was indeed a mathematical curve, which Deleuze related to continuous functions and to Leibniz's invention of differential calculus; but the early digital avant-garde preferred to interpret even the Deleuzian "fold" as an angular crease, in the tradition of Peter Eisenman's deconstructivism (and Eisenman himself was central to this part of the story).[51] In short, nothing predestined the first wave of digitally intelligent designers to become streamliners. Nothing, that is, except the ease of use of the new spline-modeling software that became available in the early 1990s.

Spline modelers are those magical instruments, now embedded in most software for computer graphics and computer-aided design, that can script free-form curves of any kind, and translate every random cluster of points, every doodle or uncertain stroke of a pencil, into perfectly smooth and curving lines. This apparently inexorable program of universal polishing of the man-made environment (which is in fact a contingent side effect of the mathematical tools at our disposal for its notation and representation) derives from a complicated genealogy of mechanical, mathematical, and computational developments, each offering a particular take on the matter: sometimes aimed at finding the smoothest line through some arbitrary points; sometimes at the design of a randomly continuous curve; sometimes at its approximation through recursive subdivisions, and so forth.[52]

The most important component of today's curve-generating software derives from studies carried out in the late 1950s and early 1960s by two French scientists, Pierre Bézier, an engineer by training, and mathematician Paul de Casteljau, working for the Renault and Citroën carmakers, respectively. A few years apart, de Casteljau and Bézier found two different ways to write down the parametric notation of a general, free-form, continuous curve (or of more such curves joined together). The two methods eventually merged, and although it is now known that de Casteljau's work came first, Bézier was first to publish his findings (as of 1966). As a result these curves are known to this day as Bézier curves; recent literature distinguishes between Bézier's curves and the de Casteljau algorithm still used to calculate them.[53] Neither Bézier nor de Casteljau used the term "spline," and there is evidence that Bézier perceived his own mathematical breakthroughs as conceptually independent from the mechanical and mathematical tradition of spline making.[54] The term "spline" derives from the technical lexicon of shipbuilding, where it described slats of woods that were bent and nailed to the timber frame of the hull.[55] The slats had to join those structural points in the smoothest possible way to avoid turbulence in the streamline (the line of contact between the water stream and the hull) and to limit drag. Later on, similar operations were performed by hand in other branches of engineering for similar aerodynamic reasons; the term equally refers to flexible rubber strips that technical draftsmen used until recently for drawing smooth curves between fixed points (drafting or draughting splines; or lofting splines when executed in full size). A spline is thus the smoothest line joining a number of fixed points, but there seems to have been no scientific definition of it before 1946, when the mathematician Isaac Jacob Schoenberg used the term to designate a new function he invented to calculate interpolations in statistical analysis. Basis Splines, or B-Splines, as these mathematical functions were then called, were eventually upgraded to include Bézier curves, and further generalized under the name of NURBS (for Non-Uniform Rational B-Splines): NURBS are today the most common notation for free-form curves in all branches of digital design and manufacturing.[56] Recursive subdivisions, another method of curve generation, can approximate and parameterize existing shapes and volumes as found, and for that reason, subdivision-based CAD software is largely used by the animation industry.[57] When animation software first became largely available, in the late 1990s, designers often saw subdivisions as an alternative, more "naturalistic" approach to free-form, unbounded by the engineering constraints of early CAD/CAM software; the mathematical premises of subdivision algorithms are also very different from those of splines and Bézier curves. Yet for the last ten years or so, "Subs," or "Sub-Ds," as students sometimes call them, have been mostly used as different means to the same end—namely, to generate smooth parametric curves and surfaces. Regardless of the mathematics used to notate them, or of the software used to draw them, splines, NURBS, Bézier curves, and Subs generate high-tech, sleek, streamlined images and objects from which each sign of human intervention—the wavy, uncertain trace of the gesture of the human hand and its analog, variable tools (angles, junctions, gaps)—have all been removed.

The original purpose and task of Bézier's and de Casteljau's math was, indeed, to eliminate precisely that kind of manual approximation from the design process. As Bézier and de Casteljau both recount in their memoirs, the mathematical notation of curves and surfaces was meant to increase precision, reduce tolerances in fabrication, and save the time and cost of making and breaking physical models and mock-ups.[58] We can see why carmakers in the 1950s and 1960s—particularly French ones—would have been interested in that technology: streamlining (aerodynamics) was then popular in car design, but the molds to cast dies for the metal presses had to be individually handmade by artisan model-makers, as were all the prototypes before production; the final design of a car, for example, was not a set of drawings but a 3-D master model made in hardwood, from which blueprints and their measurements would be derived as and when needed. The famously aerodynamic Citroën DS was designed in Paris in the years immediately preceding de Casteljau's studies—and entirely by hand; which is probably one reason why de Casteljau, a young mathematician then just back from his military service in Algeria, was hired by a research department at Citroën called "détermination mathematique des carrosseries" (mathematic determination of bodywork), headed by a Mr. Vercelli, about whom nothing more is known.[59]

Bézier's and de Casteljau's research, at the start, appears to have been primarily motivated by the mathematical ambition to translate general free-form curves into equations. Digital fabrication must have seemed a less urgent prospect back then: Bézier's first experiments in numerically controlled milling machines were abandoned in 1960, and the first computer-driven drafting machines he built from scratch in the years that followed must have seemed so unpromising, in commercial terms, that the Régie Renault, then state owned, allowed Bézier's team to publish a series of scholarly papers where the new technology was described in full.[60] Thus Bézier's research was at the basis of the CAD/CAM system that Renault kept developing and later adopted, called UNISURF, but also of competing proprietary technologies developed by other companies, such as the aircraft maker Dassault. On the other side of town, de Casteljau's team was bound to secrecy for longer. De Casteljau recently wrote that, due to a prolonged strike of wood modelers, the templates for the Citroën GS were the first to be produced entirely by machines, but it is not clear to what extent the bodywork itself was designed on screens rather than in clay and wood: the GS started production in 1970, and its design was developed throughout the 1960s.[61]

Of course, this was not an exclusively Gallic story. At the same time as de Casteljau's and Bézier's studies on free-form curves, research on B-Splines was being carried out at MIT, Boeing, at the British Aircraft Corporation, and particularly by Carl de Boor at General Motors, which developed its own CAD/CAM system in the 1960s; in 1981 Boeing was among the first adopters (or perhaps the inventor) of NURBS, which in the same year were endorsed by the Initial Graphic Exchange Standard (IGES), a consortium of industry and government bodies.[62] Yet Bézier often recounted that, as late as 1971, after one of his presentations, a manager at Renault objected, "If your system were that good, the Americans would already have invented it!"[63] In 1969, Robin Forrest, whose work on Bézier's curves would contribute to the generalization of B-Splines to NURBS, had to travel from Cambridge, England, to the GM Research Laboratories in Detroit to be shown "the crazy way Renault designs surfaces."[64] And in July 1991, when the office of Frank Gehry in Los Angeles started to look for a suitable CAD/CAM technology for the design of a building in the streamlined shape of a fish (which would become the Barcelona Fish, still prominently floating over Barcelona's Olympic Marina), through a succession of phone calls and the intercession of Bill Mitchell at MIT and of Rick Smith at IBM, Gehry's office was eventually referred to Dassault's headquarters in Paris. The role of Dassault's CATIA software in contemporary architecture is, to this day, better known than its industrial history before and after its architectural reincarnation.[65]

Few of today's designers would have become keen spline-makers if they had had to make each spline by hand, bending slats of wood, or if they had had to slog through all of Bézier's math with paper, pencils, and a slide rule. This is one reason why free-form curves were seldom built in the past, except when absolutely indispensable, as in boats or planes, or in times of great curvilinear exuberance, such as the baroque, or the Space Age in the 1950s and 1960s. But, as it happens, in one of those only apparently serendipitous encounters that mark the history of technological change, mathematics and technology here kept crossing paths. As computers became smaller and cheaper, CAD software migrated from corporate mainframe computers to workstations to desktop personal computers; AUTOCAD, the first CAD tool designed for MS-DOS, was released in 1982.[66] As of the early 1990s affordable commercial software for computer-aided design started to include powerful spline modelers that made Bézier's math easily accessible through graphic user interfaces: control points and vectors that anyone could easily edit, move, and drag on the screen with the click of a mouse. This game turned out to be faster and more intuitive than the mathematics on which it was based, and digital designers started to play it with gusto. Form-Z, the most influential of these early packages, was developed at Ohio State University, apparently with the complicity of Peter Eisenman, and released in 1991.[67]

Bézier's curves, B-Splines, and NURBS are pure mathematical objects, based for the most part on differential calculus; their smoothness, which we perceive as a visual and tactile quality when splines are built, is a quantifiable entity throughout the design and production process, defined by one or more derivatives to the function of the first curve. Mathematical objects do not fit with the phenomenological world we inhabit; designers using spline modelers "model" reality by converting it into a stripped-down mathematical script, and the continuous lines and uniform surfaces they draw or make in physical reality are ultimately only a discrete, material approximation of the mathematical functions they use to notate them and computers then use to calculate as many points belonging to them as needed.[68] Of course, not every digitally intelligent designer in the 1990s was a pure spline-maker: Greg Lynn and Bernard Cache explicitly claimed to use calculus as a primary tool of design,[69] while Frank Gehry (for example) famously used computers to scan, measure, notate, and build the irregular, nongeometric three-dimensional shapes of his handmade maquettes.[70] From its inauguration in 1997, Gehry's Guggenheim Bilbao (designed from 1991 to 1994) was hailed as a global icon of the new digitally driven architectural style, but as most digital designers at the time used curve-generating software to even out, somehow, all final lines and surfaces, the divide between free-form, subdivisions, and mathematical splines is often a tenuous one, and not easy to perceive: regardless of the process (based on mathematics from the start, as in the case of Cache and Lynn, or derived at least in part from natural accidents, as in the case of Gehry), what one sees if one just looks is, simply, a landscape of sweeping, spliny curves. This is the visual aspect that was mostly noted at the time, and defined the style for which the first digital age became famous, for better or worse.

Fast-forward to 2016. The Internet boom famously busted in 2001, but the spline-dominated, curvilinear style now often associated with the "irrational exuberance" of the digital 1990s lived on, still successfully practiced by some of the early pioneers and handed down to a new generation of younger digital designers. With technical progress, many visionary predictions from the digital 1990s are now becoming a reality. Big, streamlined surfaces can now be built at more affordable prices, and recent projects by Zaha Hadid and others deploy this sinuous language at ever bigger and bolder scales, with a level of technical and formal virtuosity that would have been unimaginable only a few years ago. Today this style is often called "parametricism,"[71] and at its core mathematical splines and NURBS are its most distinctive notational and technical tool. Mathematical splines, in turn, are based on analytic geometry and calculus: Newton's and Leibniz's calculus is, to this day, the best instrument to describe continuous lines that are characterized by variations, and variations of variations. The mathematics of free-form is thus the zenith and culmination of a historical process that started with

Descartes, Leibniz, and Newton: Baroque mathematics found a way to use equations to replace the drawings of some simple geometrical figures—straight lines and conics; using basically the same tools, with only marginal upgrades, today we can notate any curve whatsoever. Bézier, so to speak, finished the job that Descartes and Leibniz had started. Yet once again, if we look at the history of mathematics in terms of pure quantitative data analysis, it is hard not to see calculus—just like logarithms, another great invention of baroque mathematics—as another, but even more astounding small-data technology: perhaps the ultimate small-data technology of modern science.

2.6 From Calculus to Computation: The Rise and Fall of the Curve

Consider the mathematical notation of any continuous line in the well-known format $y = f(x)$. How many points does any such notation include and describe? Plenty: an infinite number of them. In practice, that script contains all the points we would ever need in order to draw or produce that line at all possible scales. But let's assume, again, to the limit and *per absurdum*, that we can have access to unlimited, zero-cost data storage and processing power. In that case, freed from the need to skimp on data, we could easily do away with any synthetic mathematical notation and record instead an inordinately long, dumb log: the list of the positions in space (x-, y-, z- coordinates) of as many points of that line as necessary. The resulting cluster of points would not appear to follow any rule or pattern, nor would it need to, so long as each point is duly identified, tagged, and registered—ready for use, so to speak. This is exactly the kind of stuff humans don't like, but computers do well. That mathematical script (the $y = f(x)$ functional notation) is a compact, economical, small-data shorthand we use to replace what is in fact an extraordinarily long list of numbers. As the list itself would be too long for our

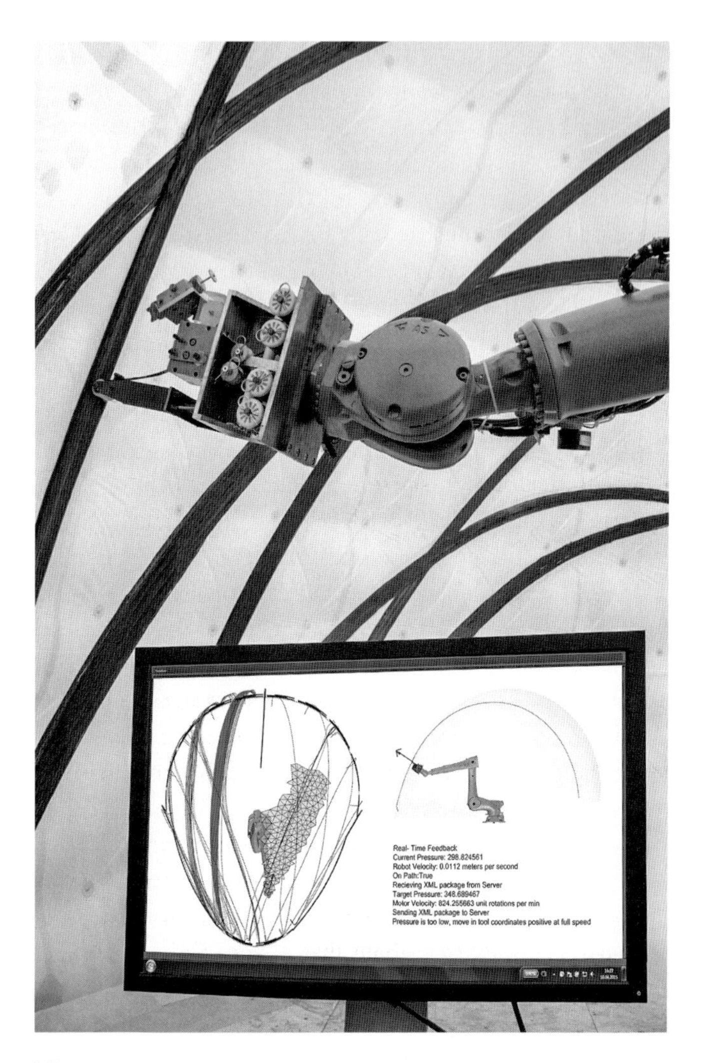

2.15

ICD Institute for Computational Design (Prof. Achim Menges). ITKE Institute of Building Structures and Structural Design (Prof. Jan Knippers). *ICD/ITKE Research Pavilion 2014–15*, sensor driven real-time robot control of cyber physical fiber placement system, University of Stuttgart, 2015. © ICD/ITKE University of Stuttgart.

small-data skills, we convert it into a very short formula. That formula is much easier for us to manipulate, process, and remember than the list itself would be; just like any of the modern classification systems discussed above, an equation or function also, miraculously, allows us to retrieve all the events it refers to (in this instance, to recalculate the coordinates of all the points it indexes), whenever needed. Thus an equation compresses, almost miraculously, an infinite number of points into a short and serviceable alphanumerical script.

That works very well for us; which is why we still celebrate the names of Descartes, Leibniz, and Newton and we still study analytic geometry and calculus at school—in fact, this is why all the math we study at school, starting from the age of six, is aimed and finalized at getting there—at mastering differential calculus more or less by the time we reach adulthood. But computers do not work that way. When fed a raw, unstructured list, log, or inventory of any size, computers can just keep it as it comes—in its pristine sequence or any other, even if random or haphazard. Computers can search and retrieve each item in any list, regardless of the way that list is or is not ordered, because this is what computers do best: unlike us, and just like Gmail, computers can search without any prior sorting.[72] That list may be unimaginably long, but computers don't care; and the computer's big data logic makes perfect economic sense, if data cost nothing. A list where raw data are kept unsorted does not make any sense to us, but computers do not care about that either. We humans need to sort (organize, classify, formalize, order, structure) a list to make it usable (so we can retrieve the items in it) and to make it meaningful (so we can organize ideas and things by hierarchies or orders of causation). Computers are not in the business of finding meanings and can use any huge, messy, untreated, and unprocessed random inventory just fine: they can search without sorting; hence they can predict without understanding. And, apparently, in many cases computers can already predict that way better than we can in our own, traditional, small-data way—which was that of modern science.

The calculus-based spline is a quintessential small-data tool. As a design figure, splines are space-age technology; they belong with the Beatles and flared jeans. Let's be honest: if we only look at forms, or style, and we forget about the technology, the digital spline of the 1990s was a revival. Computers made streamlining cheaper and better; easier to design and make—but streamlining was certainly not a new idea, and in the last decade of the twentieth century the spline was certainly not a new form. By providing computational tools and graphic user-interfaces to Bézier's math, digital spline modelers gave streamlining a new lease on life. That was a very good idea twenty years ago, when computers, processing power, and data were expensive. In that context, it made perfect sense to use computers to emulate and replicate the small-data logic of modern mathematics—in a sense, to make computers imitate us. But in today's big data environment that mimetic effort is no longer necessary—indeed, it is no longer warranted. Computers can work better by following their own logic. We can make computers sort before searching, the way we do. But computers already achieve much better results when we let them search without sorting, the way we don't and can't do.

Revolutions in manufacturing tend to happen first at a small scale, and scaling up may sometimes be late in coming. In this instance, digital splines revolutionized graphic design one decade before they changed the history of world architecture. It all started with laser printing. Mechanical printers can only print from a limited library of built-in metal fonts: think of a typewriter. The interchangeable typeballs or daisy wheels in the electric typewriters of the 1960s and 1970s allowed typists to switch between a few styles of fonts, but replacing the typeballs or wheels took time.[73] Laser printers, to the contrary, can print all kinds of fonts (and indeed any rasterized image), seamlessly and from the same machine; so, when affordable laser printers became available, in the mid-1980s, word processors started to upgrade their library of fonts, adding new styles and sizes.[74] Following the traditions of typography, each font (for example, Times) should then have been designed anew for each size (8p, 10p, 12p ...), and each glyph digitized as a rasterized map of pixels; but this would have created graphic files far too big for the limited memories and processors of early personal computers. The solution came with Adobe's PostScript software, first released in 1984.[75] PostScript notated the design of each glyph mathematically, as a combination of straight lines and Bezier curves. The advantage was that the same script would fit all sizes, because the same formula would generate the same drawing (say, the lowercase *a* in Times font) at every scale the available software and hardware would support, on the screen as well as in print, and each of these scalable signs would look the same (hence the acronym, then so popular, WYSIWYG, for "what you see [on the screen] is what you get [in print])." Thus, thanks to the math of free-form curves (Bézier's, etc.), a vast and unwieldy graphic library was compressed into a file so small that it could run on most PCs of the time (which incidentally also led to the desktop publishing revolution of the late 1980s and 1990s).

Once again, however, that entire strategy would be unwarranted in today's computational environment. Processors and storage devices are now so cheap and powerful that huge graphic libraries of all kinds can now be kept and processed almost everywhere without the need for any sophisticated compression technologies. For example: if we wanted to, we could redesign each glyph to allow for design variations specific to each size of a font (as was the case in traditional typography), and we could do that for an inordinate number of different fonts. That would create a very large inventory of bitmaps—but again, today that would hardly be a problem. Of course, no matter how big that inventory, the number of available bitmaps would always be limited, and as bitmaps are not scalable, chances are that sooner or later someone would fail to find a font in the exact size or resolution needed. Using splines that would never happen: that is indeed the great notational and, in a sense, ontological advantage of the mathematics of continuity against all arithmetics of discreteness, old and new alike: an advantage which, in this instance, we would deliberately relinquish. Would that matter? Today, we might Google the font we need and find out that it already exists somewhere. Someone could be tasked to add the missing parts of the design when the time comes. Or we might let a few pixels show in all of their coarse discreteness for a while, until someone, or something, interpolates, or fills the gaps. This is what some designers of the second digital age are now doing.

2.7 Excessive Resolution

In different ways, today's digital avant-garde has already started to use Big Data and computation to engage somehow the messy discreteness of nature as it is, in its pristine, raw state—without the mediation or the shortcut of elegant, streamlined mathematical notations. The messy point-clouds and volumetric units of design and calculation that result from these processes are today increasingly shown in their apparently disjointed and fragmentary state; and the style resulting from this mode of composition is often called voxelization, or voxelation. The Computational Chair Design Studies by Philippe Morel of EZCT Architecture & Design were among the earliest demonstrations of this approach,[76] and the 2013 ArchiLab exhibition in Orléans, France, reveals this new formal landscape at a glance (see, for example, works by Alisa Andrasek and Jose Sanchez, Marcos

Cruz and Marjan Colletti, Andrew Kudless, David Ruy and Karel Klein, Jenny Sabin, and Daniel Widrig).[77] Subdivisions-based programs, originally used to simulate continuous curves and surfaces, today are often tweaked to achieve the opposite effect, and segments or patches are left big enough for the surface to look rough or angular. Discreteness is also at the basis of the method of finite elements (seen above),[78] now embedded in most software for structural design, and which represents in many ways an early example of "agnostic" science,[79] where the prediction of structural behavior is separated from causal interpretations.

More examples could follow, but the spirit of the game is the same: in all such instances, designers use the power of today's computation to notate reality as it appears at any chosen scale, without converting it into simplified and scalable mathematical formulas or laws. The inherent discreteness of nature (which, after all, is not made of dimensionless Euclidean points or of continuous mathematical lines but of distinct chunks of matter, all the way down to molecules, atoms, electrons, etc.), is then captured and, ideally, kept as it comes, or in practice as close to its material structure as needed, with all of the apparent randomness and irregularity that will inevitably show at each scale of resolution. Evidently, the abstract continuity of the spline does not exist in nature: we can write down splines as mathematical formulas and imagine them as a seamless flow of Euclidian points, but in physical reality we can only make most of them by discrete pieces, by pixels or voxels—which can only be as small as the maximum resolution supported by the display, printer, or physical interface we are using.[80] The manufacturing tool which best interpreted the spirit of continuity of the age of spline making was the CNC milling machine, a legacy subtractive fabrication technology that, using computer-controlled drills, could at its best simulate the sweeping, smooth, and continuous gestures

2.16
Zaha Hadid Architects, *Heydar Aliyev Centre*, Baku, Azerbaijan (2007–12). Photo: © Hufton + Crow.

2.17
Philippe Morel / EZCT Architecture & Design Research, *Studies on Optimization: Computational Chair Design Using Genetic Algorithms (with Hatem Hamda and Marc Schoenauer)* (2004). The version "T1-M 860" is obtained through the optimization of 860 generations (86,000 Finite Elements Analysis–based evaluations). EZCT Architecture & Design Research © 2004. Photo: Ilse Leenders.

2.18

Alisa Andrasek and Jose Sanchez, *BLOOM,* a crowdsourced garden, urban toy, and social game, developed for the London Olympics (2012). Visitors were invited to change the layout of the initial pavilion and to add or combine new pieces for seating or other purposes. © Alisa Andrasek and Jose Sanchez.

of the hand of a skilled craftsman—a sculptor, but also a baker, or a wax modeler.[81] Not surprisingly, the CNC milling machine was the iconic tool of the 1990s, and there was a time when every school of architecture in the world had, or wanted, one. Today the 3-D printer has taken its place: an additive fabrication technology, where each voxel must be individually designed, calculated, and made.

In its present form, the distinction between subtractive and additive making goes back to Leon Battista Alberti's *De Statua* (composed in Latin at some point between 1435 and 1472), and to Michelangelo's *Letter* to Benedetto Varchi of 1549 (see chapter 3). Not surprisingly, Michelangelo thought that only the effort of taking matter away from a block of solid stone (*per forza di levare*) was worthy of the name of sculpture. In more recent times, subtractive fabrication was the first to go digital: numerically controlled milling machines have been in use since the early 1950s, first driven by punched cards and tapes, then by electronic computers (hence the term CNC, for Computer Numerically Controlled). Three-dimensional printing and additive fabrication came later: stereolithography, invented in 1984, was known and marginally available in the 1990s, but cheap and versatile 3-D printers (including desktop 3-D printers) were launched only in 2008–09,[82] and the technology has taken the forefront of digital design and culture in the last few years.[83] At the time of writing, 3-D printing is as influential for a new generation of digital makers as 3-D milling was at the change of the millennium. This is ostensibly due to a number of developments in the technologies of manufacturing, but the two processes are also based on contrary, indeed incompatible, informational logics.

CNC milling can be as data rich as one wants it to be, or the machine allows, but in the absence of any signal (that is, in the case of zero-data input) digital subtractive technologies will still work—and deliver a plain, solid chunk of matter: in most cases, releasing the original surface in its pristine material state, unmarked, and without any denting, milling, or amputation. On the contrary, in additive technologies each voxel must be individually printed; in the absence of signal, additive fabrication delivers nothing at all. Hence digital milling can make do with few data, or even with no data—and indeed designers using digital subtractive technologies apply data to inform matter only as and where needed, whereas each 3-D printed voxel requires a certain amount of data and of machine time.

Furthermore, as each voxel is individually printed, and 3-D printing does not involve any reusable cast, mold, stamp, or die, there is no need, and no incentive, to make any voxel-generated volume identical to any other, regardless of scale or size. Mechanical printing technologies are matrix based, and any matrix, once made, must be used as many times as possible to amortize its cost. But standardization does not deliver any economy of scale in a digital design and fabrication workflow: just as twenty years ago we learned that we could laser print one hundred different pages, or one hundred identical copies of the same page, at the same unit cost, today we can 3-D print any given volume of a given material at the same volumetric cost, based on the number of voxels that compose it (that is, on resolution), not on geometry or configuration (that is, regardless of where each printed voxel will be relative to all others in the same volume).[84] An economist would say that the marginal production cost of a voxel is always the same, no matter how many we print—and irrespective of how they will be assembled. Thus 3-D printing brings the logic of digital mass customization from the macro scale of product design to the micro scale of the production of physical matter—and at previously unimaginable levels of complexity and granularity: recent 3-D printers can create objects with variable densities and in multiple materials.

These simple technical truisms have remarkable consequences. Let's consider a seminal example of monumental 3-D printing, Michael Hansmeyer and Benjamin Dillenburger's now famous digital grotto, commissioned by Frac Orléans and shown there in the summer of 2013.[85] In spite of and against all appearances, the grotto was not carved from a block (in the subtractive way); it was printed from dust—that is, almost from nothing—in the additive way. As a result, the grotto we see, including all of its intricate details, was faster and cheaper to make than a plain full block of that size (if printed at the same resolution), simply because the void inside the grotto was not printed. If we had wanted the plain full block, we should have kept printing—and would have kept spending. Likewise, any onlooker familiar with the traditional (manual or mechanical, or even early digital) ways of making may instinctively assume that the astoundingly intricate detailing of the grotto must have cost even more labor and money—that is, labor and money that would have been saved had this detailing not been added. Not so: as all 260 million surfaces in this 30-billion voxel space had to be individually 3-D printed, the technical cost of delivering the same number of voxels in regular rows, so as to create plain, flat, and regular surfaces, would have been exactly the same.

This is a rather counterintuitive result, as we tend to think that decoration, or ornament, is expensive, and the more decoration we want, the more we have to pay for it. But in the case of Hansmeyer's grotto, the details and ornament we see inside the grotto, oddly, made the grotto cheaper. Of course this is only true in theory, if we disregard the time and cost of designing each voxel one by one—an operation that appears to have required some drastic design shortcuts, interpolations, and simplifications.[86] All the same, the difference in technical and theoretical terms is striking, even revolutionary. Since the beginning of modern times, indeed since Leon Battista Alberti, Western

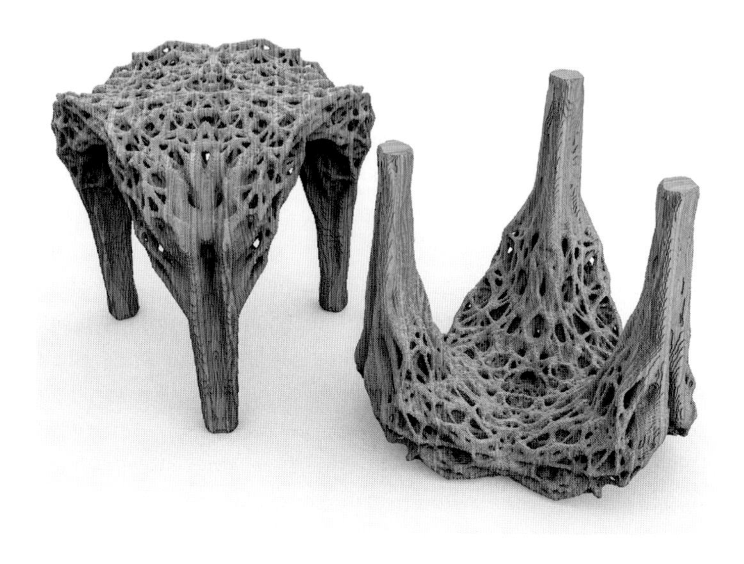

2.19
Daniel Widrig, *Degenerate Chair*, Frac Centre Collection (2012). Author's rendering.

architecture has developed a systematic theory of ornament as supplement: something which is added on top of an object or a building, and which can be taken away if necessary.[87] But for this very reason, puritans, Taylorists, and modernists of all sorts have always blamed ornament as waste, superfluity, and, in Adolf Loos's famous slogan, a crime—labor and capital thrown out the window, money that should have been better spent in some other way.

It now appears that the technical and cultural premises of all this are simply not true anymore. In the age of Big Data and 3-D printing, decoration is no longer an addition; ornament is no longer a supplemental expense; hence the very same terms of decoration and ornament, predicated as they are on the traditional Western notion of ornament as supplement and superfluity, do not apply, and perhaps we should simply discard these terms, together with the meanings they still convey.[88] This opens a Pandora's box of theoretical issues, which in turn undermine some core aesthetic and architectural principles of both the classical and modernist traditions.

2.8 The New Frontier of Alienation, and Beyond

As several critics have pointed out, the opulent detailing shown in recent 3-D printed pieces appears to require an almost inhuman—or posthuman—level of information management: no one can notate 30 billion voxels one by one without the intervention of a more or less intelligent tool for the machinic or automatic control of some design features, and these big-data works display a kind of design intelligence that is already closer to that of the machine than to ours.[89] Others have also pointed out that the visual and tactile density evoked by these pieces is more than human senses can perceive, but that allegation is probably unfounded: no matter how much design and detailing we can now put into our work, its resolution is bound to compare poorly with anything that already exists in nature—in organic nature, and even in inorganic nature.[90] Admittedly, the pieces we fabricate can now be designed at much smaller scales than the standard items of modern industrial mass production used to be (say, a steel I-beam); but in that too we are simply, little by little, getting closer to nature—not such an uncommon outcome among the mimetic arts.

Similar arguments have also been recently invoked to construe a theory of today's digital style as a style of realism, excessive realism, or digital hyperrealism. In an interesting and quirky book just published by polymath designer and theoretician Michael Young, the richness in detail and figural sumptuosity that are increasingly transparent in some work of today's digital avant-garde are interpreted as an "estrangement" device, as defined by Viktor Shklovsky in 1917 (sometimes called "distancing effect" or "alienation effect," and today better known as Bertolt Brecht's hallmark performing art technique).[91] In this instance, the distancing effect appears to derive from mostly technical factors, involving streaks and strands of Heidegger's theory of the *Unzuhandenheit* (unhandiness of the technical object), better known today among digital theoreticians through the recent interpretations of Bruno Latour and Graham Harman: some technical objects that we do not even notice when they function become perceivable and hugely meaningful when they fail, or when some symptomatic anomalies suggest the possibility of an imminent failure.[92] Michael Young cites as an example a 1999 work by the artist Jeff Wall, where what looks like a normal photograph is in fact a subtle montage of indexical incongruities acting in the background, almost surreptitiously, as an "intensifier of aesthetic attention," and suggesting that while all seems right in the picture, something, somewhere, must be quite wrong.[93] According to Michael Young, this unhandiness is a stylistic feature of today's digital avant-garde: the overwhelming richness of digitally created detail induces feelings of discomfort, or estrangement, similar to the "weird realism" that Harman famously attributes to the horror fiction writer H. P. Lovecraft (a cyberpunk and cyberpulp cult writer, as well as the subject of the first published essay by the acclaimed misanthropic writer Michel Houellebecq).[94] Assuredly, excessive resolution is a diacritical trait of the second digital style, and it has reasons to appear "weird." Excessive resolution is the outward and visible sign of an inward and invisible excess of data: a reminder of a technical logic we may master and unleash, but that we can neither replicate, emulate, nor even simply comprehend with our mind. Again, this is not entirely unprecedented: just like the Industrial Revolution created prosthetic extensions that multiplied the strength of our natural bodies, the digital revolution is now creating prosthetic extensions that multiply the strength of our natural intelligence; but just like mechanical machines did not abide by—indeed, they often subverted—the organic logic of our bodies, digital machines now do not abide by—indeed, they often subvert—the organic logic of our minds. Thus, just like industrial products embodied an artificial technical logic that went counter to that of natural handmaking (and many did not like that back then), computational products now embody an artificial logic that is counter to that of natural, organic intelligence—the mode of thinking of our mind, as expressed by the method of modern science (and many today do not like that). This may be one reason why the emergence of some inchoate form of artificial intelligence in technology and in the arts already warrants a more than robust amount of natural discomfort, and the feeling of "alienation," which originally, in Marx's critique of the Industrial Revolution, meant the industrial separation of the hands of the makers from the tools of production,[95] may just as well be applied today to the ongoing postindustrial separation of the minds of the thinkers from the tools of computation.

2.20

Michael Hansmeyer and Benjamin Dillenburger, *Digital Grotesque* (2013). Test assembly (2.20) and detail (2.21).

2.21
Michael Hansmeyer and Benjamin Dillenburger, *Digital Grotesque* (2013). Test assembly and detail.

2.22
Christian Kerez, *Incidental Space* (figures 2.22 and 2.23). Installation, Swiss Pavilion, 15th Venice Biennale of Architecture (2016). Neither designed nor scripted, Kerez's walk-in grotto was the high-resolution, calligraphic transcription, 42 times enlarged, of a cavity originally produced by a random accident inside a container the size of a shoebox. As in Frank Gehry's pioneering use of digital scanners to notate the shapes of irregular volumes in his handcrafted models, digital tools for design and production are turned into a seamless, universal 3-D pantograph, capable of capturing any accident of nature, of notating it at any level of geometrical resolution, and replicating it at any scale of material fabrication.

2.23

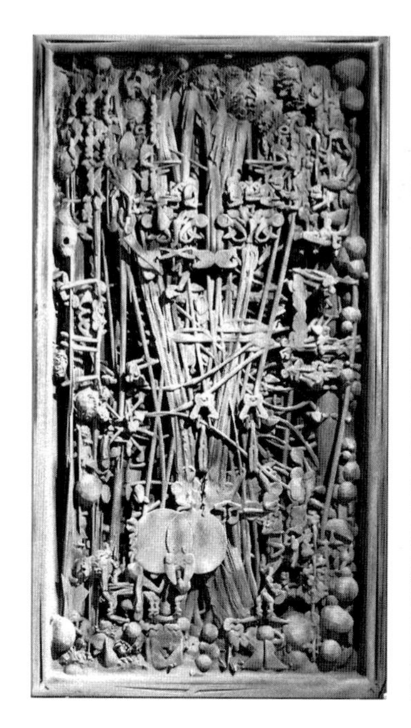

2.24
Marjan Colletti, *Plantolith* (2013), 250 kilogram silica sand 3-D print. Photo: Marjan Colletti.

2.25
Quaquaversal Centrepiece at the Spring-Summer 2016 Iris Van Herpen ready-to-wear collection (Musée d'Historie de la Médicine, Paris, October 8, 2015). Three robotic arms from the University of Innsbruck's REXILAB, dressed up as fable-like creatures, manipulate and 3-D-print actress Gwendoline Christie's dress. Design: Iris van Herpen, Jolan van der Wiel, Marjan Colletti + REXILAB. Photo: Marjan Colletti.

2.26
Young and Ayata, *Still Life with Lobster, Silver Jug, Large Berkenmeyer Fruit Bowl, Violin, Books, and Sinew Object After Pieter Claesz, 1641* (2014). Digital rendering and photomontage. Team: Emmanuel Osorno.

2.27
Young and Ayata, *Base Flowers*, Volume Gallery, Chicago (2015). Multimaterial 3-D print, resin, full color sandstone. Team: Sina Ozbudun, Isidoro Michan.

2.28
Alisa Andrasek, Wonderlab, AD Research Cluster 1, B-Pro M.Arch Architectural Design, The Bartlett UCL, *Gossamer Skin* (2016). Building skin based on environmental data (light, temperature and acoustics) and robotically 3-D printed. Tutors: Alisa Andrasek, Daghan Cam, Andy Lomas. Robotics: Feng Zhou. Projects/Students: Supanut Bunjaratravee, WeiWen Cui, Manrong Liang, Xiao Lu, Zefeng Shi. © Alisa Andrasek, AD Research Cluster 1, The Bartlett UCL.

The critique of the "weird realism" of some contemporary digital avant-garde appears to have been inspired by a recent philosophical movement known as speculative realism, or object-oriented ontology. Yet, in spite of a spate of recent publications on the matter, the two trends do not appear to have much more in common than the name—and a very general one at that. In the winter of 2015 several contributors to the journal *Log* tried to ascertain what "an object oriented architecture would look like," arguing that the philosophy of speculative realism relates to a new design sensibility based on "joints, gaps, … misalignments, and patchiness";[96] but this simply describes one of the core traits of the second digital style, on which no philosopher of that school (or of any other) has vented any opinion. However, another aspect of speculative realism may perhaps more deeply resonate with some concerns, ambitions, and expectations of today's computational design. Regardless of what it originally meant, which is irrelevant in this context, the speculative realists' notion of a "flat ontology" is often cited to endorse the view that minerals, vegetables, animals, humans, as well as technical objects, can and should be seen as ontologically equal, all endowed with the same quantum of free will.[97]

While the idea that a stone, a cat, a parsnip, and a vacuum cleaner may decide to go out together in the evening for a drink may appear to defy common sense—or at least common experience—vitalism, animism, and spiritism have a long and distinguished tradition in the West (as sciences, beliefs, and crafts, as well as in witchcraft, sorcery, and magic). The so-called postmodern sciences of indeterminacy and nonlinearity always had a strong spiritualistic component (alongside a more positivistic, hard-science one); these doctrines and beliefs have always been a powerful source of inspiration for digital designers, particularly in the 1990s, when many thought that computers, and the Internet, would vindicate a long-standing, nondeterministic view of science and of nature. Likewise, the theories of emergence and of self-organizing systems, which have played an equally powerful role in the history of digitally intelligent architecture, always lent themselves to vitalistic interpretations—alongside more practical, instrumental ones.[98] In the end, although few would admit it verbatim, the very same notion of artificial intelligence—of an inorganic life created by humans—could vindicate many time-honored assumptions of white and black magic, and fulfill some of its objectives. Without going to such extremes, a respectable, scholarly, university-grade philosophical school claiming that all matter is equally alive—including inorganic matter, which designers craft and mold and animate, in a metaphorical sense—is likely to appeal to contemporary designers and theoreticians who may actually believe in the animation of the inorganic, as some do. Likewise, today's computer-based science of data, dealing as it does with previously unimaginable degrees of complexity, may appear as the next avatar of choice for the timeless human ambition to reach for something beyond the grasp of the human mind. As argued above, today's computation is certainly out of kilter with the methods of modern science and the processes of our mind, both hardwired for small data. But does the new science of data warrant, encourage, or even just admit of any continuing belief in the indeterminacy (if not the animation) of the natural and social phenomena it describes?

Stephen Wolfram received a PhD in theoretical physics at the age of twenty, and two years later, in 1981, he was one of the first recipients of a MacArthur "genius grant." In 1986 he started to develop *Mathematica*, a new scientific software based on the operations and functions of modern mathematics.[99] In a sense, *Mathematica* (at the time of this writing, a global industry standard for all kind of applications, and not only in science and technology) allows computers to imitate and reenact the modern science of math in its entirety—including notations, symbols, formalisms, and logic. Then Wolfram had another idea: he thought that, rather than making computers imitate us, we would be better off to let them work in their own way. He turned to cellular automata, a discrete mathematical model that had been known since the 1940s and had gained popularity in some postmodern circles in the 1970s. In 2002 Wolfram published a 1,280-page book, *A New Kind of Science*, claiming that, using cellular automata, machines can simulate events that modern mathematics cannot calculate, and modern science cannot predict.[100]

Cellular automata are rules or algorithms for very simple operations that computers can easily repeat an extraordinary number of times. Ostensibly, this is the opposite of the human logic: as human operations are slow and human time is limited, we generally prefer to go the other way, and human science takes a lot of time to develop, hone, and refine a few very general laws that, when put to task, can easily lead to calculable results. Computers, having no knack for abstraction, prefer to repeat the same dumb operation almost ad infinitum (which they can do very fast) until something happens. By letting computers do just that, Wolfram's "new kind of science" can already predict complex natural phenomena (such as the growth of crystals, the formation of snowflakes, the propagation of cracks, or turbulences in fluid flow) that modern science has traditionally seen as indeterminable or not calculable. When the right initial rules are intuited and let loose, computers will replicate those unpredictable events and play them out just as they unfold in nature. Wolfram never explained how to intuit the right rules—a process that may not be very dissimilar from that of arriving at the right causal laws in modern inferential science. As simulated tests are much faster than real ones, however, trial and error in this instance may be a viable heuristic method, and indeed Wolfram's new scientific method is very similar to the process of computational simulation already current in structural engineering: when structures are too complex to calculate or even to understand using the rules and laws of classical mechanics, engineers simulate a vast number of almost random variations and try them out in simulation (in fact, they break them on the screen), until they find one that is good enough, or does not break. Likewise, if one lets a machine try plenty of them out for as long as it takes, at some point one of Wolfram's cellular automata may strike the right sequence and replicate a complex natural process in full. Which cellular automaton will do the magic? No one can tell at the start. And why did that cellular automaton work, and not another? No one can tell at the end.

For many empiricist, positivist, and utilitarian thinkers, the primary purpose of science always was to predict the future—our understanding of nature was little more than an additional bonus, in a sense, and an ancillary means to this end. Computers can now predict things that science cannot explain. Yet, if prediction without causation always had some sulphurous zing and magic aura about it, there is not much that is magic in cellular automata. The same cellular automaton script, when rerun, will always generate exactly the same sequence, no matter how long we let it run; if one of these sequences replicates at some point some formerly indeterminable natural processes, then the natural phenomena thus described must be equally replicable. Computational simulation and cellular automata do not expand, and certainly do not extol, the ambit of scientific indeterminacy; on the contrary, they lessen and lower it, because they offer predictions where modern science didn't and doesn't. What some still call indeterminism is in fact a new kind of computational hyper-determinism—quite simply, determinism we do not understand. We use it because it works. And, evidently, we can try to make some effort to figure out how and why computers can do that.

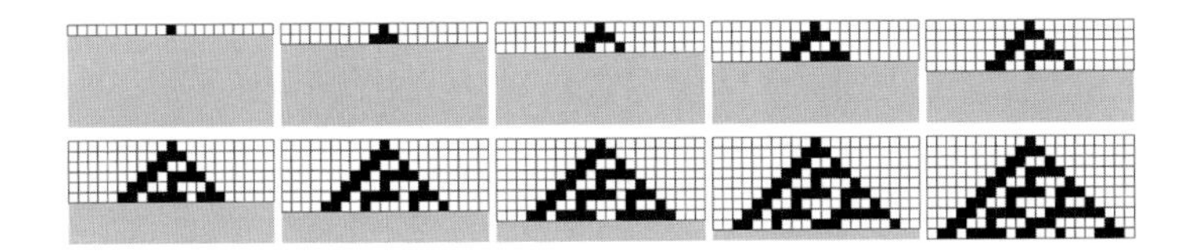

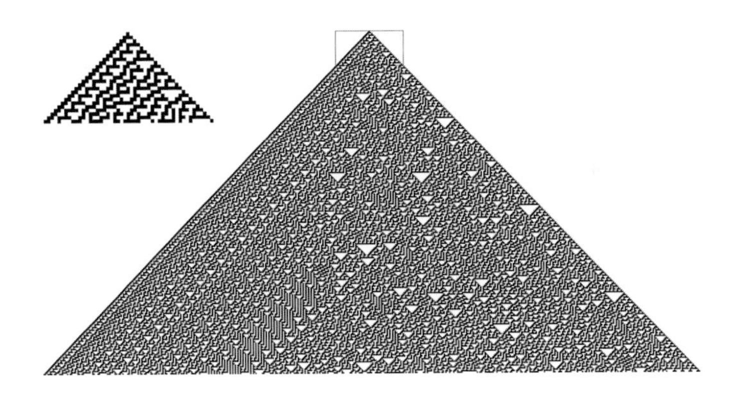

2.29
Stephen Wolfram, cellular automaton number 30, at 10, 25, and 250 steps, from Wolfram, *A New Kind of Science* (Champaign, IL: Wolfram Media, 2002), 27–29 (redesigned by A. Vougia). © Stephen Wolfram, LLC.

As I have argued, this posthuman logic is already ubiquitous in our daily lives and embedded in many technologies we use. Nothing represents the spirit and the letter of this new computational environment more than the "search, don't sort" Google tagline: humans must do a lot of sorting (call it classifications, abstraction, formalization, inferential method, inductive and experimental science, causal laws and laws of nature, mathematical formulas, etc.) in order to find and retrieve things—both literally and figuratively. Computers are just the opposite: they can search without sorting. Humans need a lot of sorting because they can manage only few data at a time; computers need less sorting—or, indeed, no sorting—because they can manage way more data at all times. To sort, humans must have a view of the world—regardless of which one came first, the worldview or the will to sort; computers need neither.

Architects adopted digital technologies earlier, and more wholeheartedly, than any other trade, profession, or craft. Since the early 1990s the design professions have been at the forefront of digital innovation, and many key principles of the digital revolution—from digital mass customization to distributed 3-D printing—have been interpreted, developed, popularized, if not outright invented by architects and designers. Wolfram's cellular automata are now the hottest topic in schools of architecture around the world, where students use them to do all kind of things—just as they did with spline modelers twenty years ago. Now, like then, the use and abuse of these tools leave a visible trace—and that shows in the style of the things we make. But digital designers, due to their position as trendsetters and early adopters, are also ideally positioned to capture, interpret, and give visible form to the technologies they use. When Frank Gehry's Guggenheim Bilbao was inaugurated in 1997, it was immediately recognized as the emblem of a new way of building—and of a new technical age. That building—and a few similar ones, but none with the same force of persuasion—proved to the world that, using digital technologies, we could realize objects that, until a few years before, few architects could have conceived, and no engineer could have built. And if new ideas and new technologies can to such extent revolutionize building, one may be led to think that the same may be true of any other field of art and industry and of human society in general. Such is the power of persuasion that architectural forms can wield in some fatidic moments of architectural history. The first digital turn was one such moment. For better or worse, the second digital turn now unfolding may be another.

3 THE END OF THE PROJECTED IMAGE

From futurism and cubism to expressionism and architectural deconstructivism, perspectival images were proclaimed dead many times over in the course of the twentieth century. Modernist avant-gardes may have had few things in common, but they were united in their aversion to Renaissance perspective and to the perspectival view of the world. Yet, due to the domination of optical, mechanical, and then digital technologies for the creation of perspectival images, still or moving, perspective remained the dominant paradigm of our visual culture, and a staple of our culture at large, until very recently. Today, at long last, the demise of projected images may be happening for good—this time around, however, not by proclamation, but by sheer technological obsolescence. A major upheaval of our visual, cultural, and technical environment is taking shape, comparable to that which was brought about by the invention of perspectival and then photographic images. Indeed, these are the primary, albeit not the only, kinds of projected images that are being phased out by the rise of digital technologies for 3-D scanning and 3-D printing; parallel projections, long favored by the technical and design professions, have already been mostly replaced by computer-based 3-D modeling.

In the brief span of less than one generation, digital technologies have moved from word processing to image processing to 3-D processing—from verbal to visual to spatial operations. This is due in the first place to steady technical advancement: words use less data than pictures and pictures use less data than three-dimensional models; as computers have grown ever more powerful and cheaper, they have been able to take on bigger and bigger data. Today, for most practical applications, the marginal cost of advanced computation is close to zero. As discussed at length in the preceding chapter, one of the consequences of the ubiquity and affordability of processing power is that there is now less urgency to laboriously cull, select, and compress data to make information leaner and easier to deal with. As a result, many technologies for data compression that were developed during the early days of computing, and more generally throughout the history of cultural technologies, are being dropped or shelved. Projected images are a case in point.

Alphabetical writing records the infinite modulations of the human voice using a very limited number of standard graphic signs. Alphabetical files are data-light: a typed page contains approximately two kilobytes of data, which is more or less the amount of data that Cicero could have inscribed on a wax tablet when taking notes in the Roman senate. The same page, if recorded as a photographic picture in coarse black or white (binary) pixels, would weigh approximately 1,000 kilobytes, or five hundred times its alphabetic equivalent. This incidentally disproves the old (but in fact very modern) saying that a picture is worth a thousand words: in purely quantitative terms the opposite is true, and in the example I gave, each word is worth approximately 3,000 binary digits (bits) of images. This is one reason why, in the early days of electronics, when processing and storage power were still rare and costly, computers recorded each letter of the alphabet as such (translated into a number).[1] But the difference in cost between the storage of a slim alphabetical file and that of the same page recorded as a pictorial image is now practically irrelevant; in fact many textual or archival databases now routinely keep both. Alphabetical files still offer plenty of advantages (for example, they are searchable for words), but we no longer need them to compress oral or visual data due to technical limitations or cost. Today, we can digitally record and process sound as sound and images as images, as need be.

Just as the alphabet was extraordinarily successful in converting an infinite number of sounds into a limited register of signs drawn on a flat surface, so perspectival images were extraordinarily successful in converting infinite distances in space into a limited register of points and lines drawn on a flat, measurable picture plane. There is no easy mathematical way to quantify the advantage of data compression in this instance, because the number of points that exist in space behind the picture plane in a perspectival construction is infinite to the power of three, while the number of points on the picture plane itself is infinite to the power of two. In practice, however, and from its early modern beginnings, the great idea behind perspective was that the points and distances behind the perspectival screen could be, in Leon Battista Alberti's cautious wording, "almost infinite,"[2] whereas the picture plane upon which this infinity is geometrically projected is not: the painting is drawn on a finite and measurable surface. A tool that can convert infinite distances in space into a measurable notation on paper is indeed a great technology for data compression, which is one reason why perspectival images have been used by all and for all kinds of purposes, from their invention to this day—regardless of the acerbic and at times jealous hostility of so many avant-garde artists.

Today, however, technologies already exist that can almost instantly record the spatial configuration of three-dimensional objects in space by notating as many points as needed in precise x, y, z (spatial) coordinates. This may not yet apply to landscapes or vast distances in open air,[3] but it does to most interior spaces and full-round objects at the size of human bodies, statuary, or even buildings. In all such instances, a 3-D scan can now be taken almost as easily as a perspectival (photographic) snapshot, and the resulting file can be saved, edited, and processed almost as easily. A 3-D scan can evidently capture much more information than any perspectival picture ever could—and today, at almost the same cost. It is easy to predict that soon and almost inevitably, as has occurred many times before in the history of cultural technologies, *ceci tuera cela*. It is equally easy to predict that the idea of the superiority of two-dimensional over three-dimensional copies, which is so ingrained in Western culture, may linger for some time—if for no other reason than old habits die hard. This idea was invented—indeed, created, almost out of the blue—by a handful of Renaissance artists, writers, and scientists. It was at the time a great, modern, revolutionary idea, and it was perfectly justified—back then.

3.1 Verbal to Visual

The rapid progress of contemporary digital technologies from verbal to visual to spatial media in the course of the last thirty years curiously reenacts, in a telescoped timeline, the entire development of Western cultural technologies. Before the Renaissance, the main vehicle for the recording and transmission of visual data was verbal, not visual: images were described using words; written words were forwarded in space and time, images were not. The last encyclopedist of classical culture, Isidore of Seville, famously epitomized the ancient mistrust for all forms of visual communication in a few memorable lines: images are always deceitful, never reliable, and never true to reality.[4] From a modern point of view, we could easily find two main reasons for that: first, although some classical painters appear to have been very good at copying nature, classical antiquity bequeathed, and likely knew, no geometrical rules for making pictures—rules whereby every painter could make the same copy of the same object, and each viewer could extract the same data from the same drawing. Second, in the absence of identical reproductions in print, all manual copies of drawings were at the mercy of the talent and good will of each individual illuminator, miniaturist, or draftsman. These two conditions combined hindered the use of images for most practical and even artistic purposes; in the absence of reproducible images, most classical books had no images at all, and the illustration of works on science or technology was limited to a handful of geometrical diagrams.[5] As Richard Krautheimer pointed out long ago, in the Middle Ages even imitation in the visual arts was "almost emphatically non-visual," and works of art were often known and reproduced from verbal, not visual descriptions.[6]

All this changed dramatically in the fifteenth century, due to the concomitant rise of two image-making technologies: perspective and xylography. Alberti famously defined perspectival images as the trace left on a picture plane by a pyramid of visual rays intersecting it: Alberti's images are prints made by light—that is, by nature itself. In Alberti's theory, the painter must choose the point of view and the direction of the central ray of the perspectival construction, but when that is done, each point of the drawing is mathematically determined: given the same geometrical conditions, every painter following the rules of Alberti's perspective will make the same drawing. When that drawing is made, each point of the three-dimensional object being drawn will translate into one point on the picture plane, and the other way around—with some exceptions, as in all projections.[7] With those exceptions made, Alberti's technology can univocally convert three-dimensional objects in space into planar notations, and in reverse, from the perspectival notation it is theoretically possible to reconstruct most visible proportions of the original object (and with some luck, or one or two additional measurements, its actual dimensions, too). That was the magic of Alberti's construction and its capital technological and ideological breakthrough: yes, perspectival images look more or less like the things we see. But, by the way they are made, they also measure whatever they portray.

At almost the same time, these new kinds of mathematical images began to be mechanically printed from woodblocks. As each woodblock generates more or less identical printed copies, the indexical logic of print multiplies the indexical objectivity of perspectival drawings. From their capture to their dissemination, modern images thus acquired a double indexical guarantee of trustworthiness: true to nature when drawn by the artist; true to the artist's drawing when reproduced by the printer.[8] At long last, these were images one could trust and use. And most people did. After so many centuries of undisputed dominion of the word, in the Renaissance the West simply went visual—and it has remained so, mostly, to this day.

3.2 Visual to Spatial

Perspectival images were so effective in emulating the three dimensions of nature that soon, according to a well-known historiographical interpretation, modern visual knowledge phased out the classical primacy of direct, tactile experience: as we learned to trust images instead of the objects they represent, the eye replaced the hand in the hierarchy of senses (and of knowledge).[9] This shift from tactile to visual values is hinted at, curiously, and almost surreptitiously, in a brief passage at the beginning of the second book of Alberti's *On Painting*.[10] Both Pliny and Quintilian had cited projections as the natural phenomenon at the origin of the art of painting: according to the classical topos they relayed, the first painters only traced, then copied the contour of a human shadow projected by sunlight onto a wall (and, Pliny added, a similar projection is at the origin of sculpture, too).[11] Alberti acknowledges but dismisses these precedents, saying he would rather side with "the poets" who think the first painter was Narcissus, beholding his own image reflected in water. But Alberti does not name his sources, and as none have been found to date, that idea may well have been his own: Alberti, who first formalized painting as a mathematical theory of projections, sees the first painting as what today we would call a selfie, and in geometrical terms a very particular projection, now known as specular reflection.[12] As we know from Ovid's myth, however, Narcissus's ur-selfie was fleeting and treacherous, and it would vanish as soon as the self-admiring youth would try to touch or kiss the surface of the water. Just like the perspectival images projected on Alberti's picture plane, or veil, Narcissus's reflection is meant to be seen, not touched. Touch it, and the spell is over—as Narcissus learned the hard way.[13]

With tactility out of the game, three-dimensionality, and its replication in sculpture and other plastic arts, would soon find itself in a weakened, almost ancillary position in the new hierarchy of early modern art. Alberti, who also wrote a treatise on sculpture, nonetheless admits his preference for painting.[14] And following Alberti, for at least two centuries and with few exceptions, artists and writers on the arts celebrated the primacy of "painting" (whereby they meant perspectival images)[15] over all other forms of communication. Painting came to be seen as equal to the written word, even competing with poetry. And on another front, the internecine strife for primacy among the new "arts of drawing," almost every person of taste and culture in early modern Europe agreed that images could replicate nature much better than any three-dimensional (sculptural) copy.

The battle of painting against poetry was an easy one, and largely consensual. Horace's famous saw, "ut pictura poesis," in its original anecdotal context meant what it still literally means: namely, that poetry is as good as painting. Yet, as Rensselaer W. Lee argutely noted long ago, Renaissance writers seemed to read the simile in reverse, and take it to mean that painting can be as good as poetry.[16] They had a point: the painters they knew were no longer lowly artisans, and painting had at long last become a fine art, or a liberal art—almost as fine and liberal as poetry had always been. No one in the Renaissance would have disputed that. To prove that painting was a better art than sculpture (and painters better artists than sculptors) was, however, a trickier matter, and the dispute, then known as the *paragone* between the arts, culminated around the mid-sixteenth century, when the Florentine humanist and historiographer Benedetto Varchi (1503–1565) posted a call for papers on the subject—Panofsky called it "the earliest public opinion poll"—and then published the replies he received in a volume, preceded by his own lengthy essay.[17] Michelangelo, the only contributor Varchi cites on the title page, not surprisingly championed sculpture. As always, Michelangelo was going against the stream.

From today's media-savvy point of view, it is easy to infer that the rise of the new print technologies in early modern Europe gave visual communication a winning edge against less easily reproducible media, including sculpture. Yet Renaissance writers on the arts do not appear to have had any awareness of, nor interest in, the reproducibility of images: the *paragone* was a scholarly dispute on the nature, functioning, and mimetic efficacy of unique, nonreproducible works of art, and only incidentally an assessment of the tools and processes that artists employed for that unique creation. The main arguments in favor of painting were set forth by Leonardo at the close of the fifteenth century, in several passages now found in the Codex Urbinas Latinus 1270.[18] Painting is more powerful than sculpture because it can represent all distances and all materials, including transparent ones, such as water, glass, or clouds; painters imitate the colors of nature, whereas, Leonardo claims—and it was universally assumed in the Renaissance—sculptors would not. Painting does not require physical exertions and painters can dress neatly and elegantly when they work, whereas sculptors toil and sweat like the worst "mechanical" artisans, their faces and bodies always caked in marble dust, like bakers covered with flour.[19] Yet even Leonardo must admit that sculpture is closer to nature when reproducing freestanding bodies.

Evidently, sculpture can reproduce a full-round object in all of its three dimensions, as we would say today, and a three-dimensional model contains much more information than any single planar projection of the same. But on this too Leonardo offers a counterargument: full-round sculpture, he claims, only provides twice as much information as perspectival drawing, as two perspectival views from two opposite vantage points (typically, front and back) show the same as a full-round sculpture. This statement was as ungenerous as it was disingenuous, as Leonardo must have known that it takes more than two drawings to make a head or bust, for example (and indeed he says so elsewhere).[20] Starting with Lorenzo Lotto, Renaissance painters sometimes provided full identification of their subjects by combining three, not two, views in the same a painting. Lotto appears to have rotated his model by approximately one hundred and twenty degrees at a time, thus offering a partial view from the back;[21] when Van Dyck was commissioned to draw a full-round view of Charles I's head to ship to Rome so Bernini could make a bust of the king without traveling, he represented Charles I in a neat architectural combination of front view, side view, and one view at forty-five degrees.[22] Why Philippe de Champaigne, given a similar utilitarian commission, represented Richelieu's bust at a slight angle and then in two identical specular profiles is not clear, and in purely notational terms it seems somewhat a waste: Richelieu's prominently gibbous nose looks exactly the same from both sides.[23]

3.1

Lorenzo Lotto, *Portrait of a Goldsmith in Three Positions,* c. 1530. Kunsthistorisches Museum, Vienna. © KHM-Museumsverband.

3.2
Anthony van Dyck, *Charles I*, 1635–36. Windsor Castle, Royal Collections. Royal Collection Trust / © Her Majesty Queen Elizabeth II 2016.

3.3
Philippe de Champaigne and Studio, *Triple Portrait of Cardinal de Richelieu*, probably 1642. National Gallery, London. © The National Gallery, London. Presented by Sir Augustus Wollaston Franks, 1869.

3.3 The Technical and Cognitive Primacy of Flatness in Early Modern Art and Science

Around 1492, Leonardo had another, stronger argument to advocate the primacy of projected images over three-dimensional copies. Full-round sculptures are seen in perspective by whoever beholds them, he claimed, without any merit of their sculptors; but paintings are put into perspective by their painters (actually, into two perspectives, Leonardo famously claimed—one made with lines, the other with colors). This the painter does by "science," "marvelous artifice," and "a very subtle investigation of mathematical studies," so that the final drawing is a "demonstration" where all proportions and foreshortening derive from the "laws" of perspective.[24] Leonardo should thus be credited with first claiming scientific precision, rather than realism, as the main strength and advantage of painting over sculpture. Indeed, who could claim that a drawing on paper is closer to its three-dimensional original than an identical copy of the original itself—with all of its appearances (except perhaps for color) and actual dimensions in space? As the elegant writer (and Raphael's occasional ghostwriter) Baldesar Castiglione (1478–1529) would point out only a few years later, that would be the same as claiming that the copy is better than the original—or, in this instance, that a picture contains more information than the three-dimensional original it represents.[25] And rightly so, Castiglione concluded: because reality is what it is, but a perspectival drawing represents (in a sense, reenacts) reality through the laws of perspective, hence in a perspectival drawing all we see is "measured."[26] Today we would say that perspectival drawings are a form of "augmented reality": they embed the proportional measurements of what they show.

Throughout the sixteenth century and beyond, the Albertian definition of painting as an "entirely mathematical"[27] construction—of perspectival images as a tool of quantification—will be the main argument invoked by the advocates of painting in the dispute between the arts. If tested for realism alone, sculpture would have easily won. Indeed, it would take one of the greatest scientists of all times, Galileo, to prove that planar drawings, regardless of their scientific superiority, are also—against all appearances—closer to nature than three-dimensional copies. Shortly after the publication of *Sidereus Nuncius* (1610), which included his famous drawings of the surface of the moon, Galileo was invited by his close friend, the painter Ludovico Cigoli, to take sides in the still ongoing *paragone* between the arts. Galileo knew perspective well, and famously used geometrical projections to calculate the height of the lunar mountains and the position of the sunspots. Yet in this instance Galileo did not insist on perspective as a scientific, measuring tool; instead, he based his essay-length reply to Cigoli on a skillful and original amplification of the topos of the superior "artificiality" of painting as an imitative art ("artificiosissima imitazione"): sculpture imitates nature while retaining its natural, three-dimensional measurements, whereas painting does so in an artificial, man-made format (by projecting volumes on a plane). Thus Galileo can, paradoxically, praise perspectival images for their higher degree of realism: projected images are in fact closer to nature than sculpture—not closer to nature as it is, in three dimensions, but closer to nature as we see it, through planar images that take shape in our eyes.

It is noteworthy that the theory of perspective, in the Albertian tradition, had never before taken into account the physiology of human sight. Indeed, Alberti had explicitly claimed that, in order to describe the geometry of the perspectival construction as it happens outside of the eye, he had to make abstraction from all other physical and physiological aspects of vision.[28] Galileo does not follow Alberti. Sculpture, Galileo claims, is a faulty mode of imitation, for it provides spatial information our eyes don't need. As human eyesight cannot see through solid bodies, three-dimensional data are always lost on the eye. Sculpture may simulate depth through shade and shadows, just as painting does, but not by being three-dimensional, because no eye can see in depth. Galileo stops short of claiming that sculpture cheats the eye (an argument more often associated with painting), but his repudiation of three-dimensional imitation is as trenchant as it is definitive: since the human eye can only see the world exactly as notated in a monocular perspectival construction, all visual representation of depth above and beyond perspectival foreshortening (and its shadowing) is unnecessary, or worse.[29]

We now know that Galileo in this instance was wrong, even though this was not the main error for which he would be blamed during his lifetime (and beyond). From science and engineering to consumer electronics, stereoscopy is now ubiquitous (and has been for a while, always tantalizingly on the verge of breaking through), and today everyone knows that our vision is based on two slightly different retinal images. Just as surveyors and topographers have always used triangulations (alignments from two vantage points) to calculate distances, our mind calculates distances based on the discrepancies between two monocular perspectival views. The results are not shown as actual measurements or numbers on a screen, the way a mathematician or a computer would do, but through the cognitive rendering of images in relief—as this is the way our mind works. Yet, strange as it may seem, stereoscopy was discovered only in 1838 by the Victorian inventor Charles Wheatstone, today better known for his contributions to electrical technology, cryptography, and telegraphy.[30] In his first groundbreaking and truly astounding paper on the subject, the great scientist (a self-taught artisan of humble birth) soberly noted that the phenomenon of "binocular vision," whereby the human mind creates "the most vivid belief of the solidity of an object,"[31] albeit possibly adumbrated by Leonardo in a passage of the Codex Urbinas,[32] was entirely neglected by all subsequent studies on vision. In fact, the possibility of any cognitive perception of distances beyond that offered by planar perspectival projections had been bluntly negated by none other than the founder of modern science.

Two centuries after Alberti's invention of perspectival images and their meteoric rise to prominence due to their mimetic qualities, their mensural reliability, and their easy reproducibility in print, the new culture and technology of planar projections came full circle, so to speak, when Galileo proclaimed planar projections to be the true and only notation of our cognitive experience of the physical world. Monocular perspectival images are identical to nature because that's the way we see things, Galileo claimed. Anyone who ever tried to pass a thread through the eye of a needle with one eye intermittently open and closed could easily attest to the opposite, yet until very recently it appears that nobody did: with Galileo's cognitive endorsement of perspectival images, the world went flat. That was a strange destiny for a technology of vision originally meant to emulate the visual perception of distances in space: in the end, the planarity of the perspectival medium became its main message—an idea that has famously held sway over modern Western art history and criticism to this day.[33] After Galileo authoritatively put it to rest, the notion of an experiential, extra-perspectival cognition of space would only be revived in the nineteenth century by a scientist of genius who could probably discover stereoscopy precisely because he was unacquainted with the history and theory of Western art. But Wheatstone's invention of "solid images" was almost exactly coeval to that of argentic photography, which gave perspectival images an extraordinary new lease on life, and made the Albertian paradigm of vision even more dominant than it had ever been. In spite of its artistic, scientific, and even medical significance, stereoscopy thus remained a marginal image-making technology, until recently confined mostly to toys, fairs, shows, and other extravaganzas.

3.4 The Underdogs: Early Alternatives to Perspectival Projections

Alberti was the chief inventor of perspectival images, but he should not be blamed for the subsequent hegemony of the perspectival paradigm. True, perspective ended up being one of the most successful of his ideas, but at the beginning it was only one in a panoply of many complementary media technologies that Alberti had designed to record and transmit the place and shape of three-dimensional objects in space. Perspective was meant to be the primary notational instrument for painters; for architects and sculptors Alberti devised other, more suitable tools.

Alberti must have been fully aware of the mensural limits of central projections. The first known attempt to reverse a perspectival image to extract some of the metrics it contains—an operation known today as photogrammetry—was made by Pietro Accolti, another Florentine Academician and a contemporary of Galileo.[34] The procedure was of limited practical interest then, as it presupposed a perspectival drawing obtained through nongeometrical means (for example, using Alberti's veil or Dürer's window); photogrammetry would become more valuable after the invention of photography, which provides a reliable, machine-made perspectival image. Photogrammetry is still taught as a branch of descriptive geometry, but it is a notoriously aleatoric and difficult operation—albeit today much facilitated by computers.[35]

Technical drawings evidently require an easier and more direct way to notate spatial measurements. In a famous passage of his treatise *On Building*, Alberti recommends that architects should avoid perspective and use instead other kinds

86

ghiamo eſſere, come F G, Dico ora, che non ſarà difficile rappreſentare ſotto gl'occhi altrui, qual ſia la reale legittima preciſa forma, & figura del ſudetto Forte, & appreſſo ancora, quanta ſia la cortina, di muraglia, che ſi diſtende tra ciaſcuno baloardo, & quanto ogni ſuo Angolo, mediante la dottrina, che ſe ne ſoggiugnerà: Et quanto al primo: Dal punto di concorſo, ò liuello F, tiriſi vna retta linea paſſante per eſempio per la eſtremità dell'angolo, ò baloardo C, & producaſi fin tanto, che peruenga ſu la linea del piano in H. Et appreſſo poi dal punto della lontananza G, ſi tiri vna retta linea (paſſante per il medeſimo angolo C,) prodotta ſimilmente fino, che peruenga ſul medeſimo piano in punto I, Soggiungo adeſſo, che ſe in punto H, muoueremo vn piombo (a beneplacito) H M, & preſa con il compaſſo la quantità del piano H I la collocheremo nella porzione N L, che in detto punto L, ci deue apparire per pianta l'angolo c. di noſtra diſegnata veduta. Ne daremo vn'altro eſempio per tanto più prati-

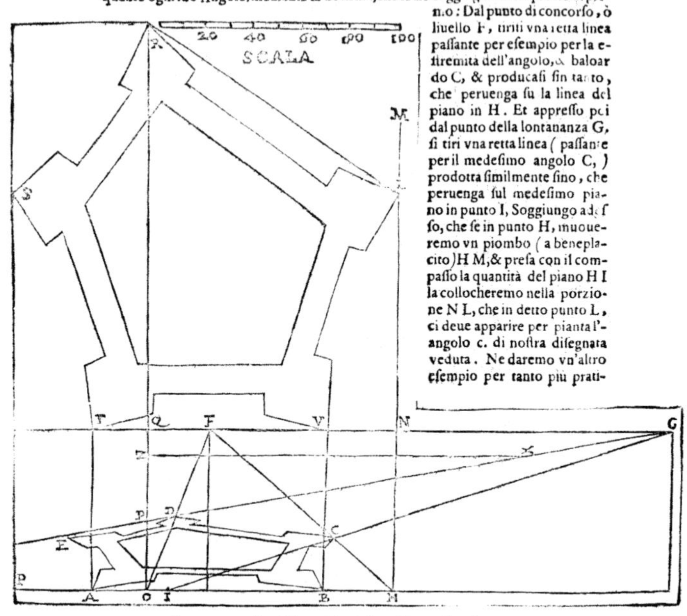

care queſta nuoua maniera di leuar Piante. Ponghiamo di voler conoſcere la punta del balloardo D. Dunque per eſſo muouaſi dal punto F, la linea F D, & producaſi in O, ſucceſſiuamente l'iſteſſo ſi faccia dal punto G, tirando la linea G D, & producendola tant'oltre, che peruēga à raſſegnarſi ancora lei ſul medeſimo piano il che non può notarſi, ò oſſeruarſi da noi per anguſtia di queſto breue ſpazio, ma conſeguiremo ad ogni modo in ſimili caſi, & accidenti l'intento noſtro, & propoſito operando in queſta infraſcritta maniera facciaſi O P, eguale à P Z, & dal punto Z, ſi tiri vna retta linea paralleia al

piano

3.4

Pietro Accolti, *Lo inganno de gl'occhi. Prospettiva Pratica* (Florence: Cecconcelli, 1625), 86.

of nonforeshortened, scaled drawings, similar to what modern designers would call parallel projections in plans, elevations, and side views.[36] Mathematically formalized by Gaspard Monge only at the end of the eighteenth century, parallel projections remained the primary notational tool of all design professions almost to this day.[37] Monge's method used two sets of parallel projections to univocally notate the position of any point in space onto two planes that, if needed, can be drawn on the same sheet of paper: descriptive geometry is a brilliant mathematical invention. When put to practical tasks, its efficacy is formidable: no one could store the Seagram Building in reality—it is quite a big building—but many offices could store (and some did, in fact, store in a few drawers) the batch of drawings necessary to make it, and, if needed, to remake it. With parallel projections (and axonometric views, which likewise were long used empirically before their mathematical rules were set forth in the early nineteenth century),[38] the art of compressing big 3-D objects onto small flat sheets of paper (or parchment or canvas or Mylar) reached the apex of modern quantitative precision: parallel projections do not even try to look like the objects they represent, but they aim at recording and transmitting the measurements of their volumes in space as precisely and economically as possible. Crucial until recently, such data cheapness is increasingly unwarranted today: using digital technologies we can already keep not only a huge number of planar drawings but also full 3-D avatars of buildings on a single memory chip—including all the data we need to simulate that building in virtual reality, or to build it in full. Oddly, even this latest technological leap had been anticipated by Alberti himself.

In his treatise *On Sculpture*, Alberti had introduced a revolutionary 3-D design and fabrication method, entirely based on digital data, to the exclusion of all drawings. Using a measuring device centered above the top of the body to be measured,

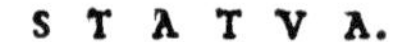

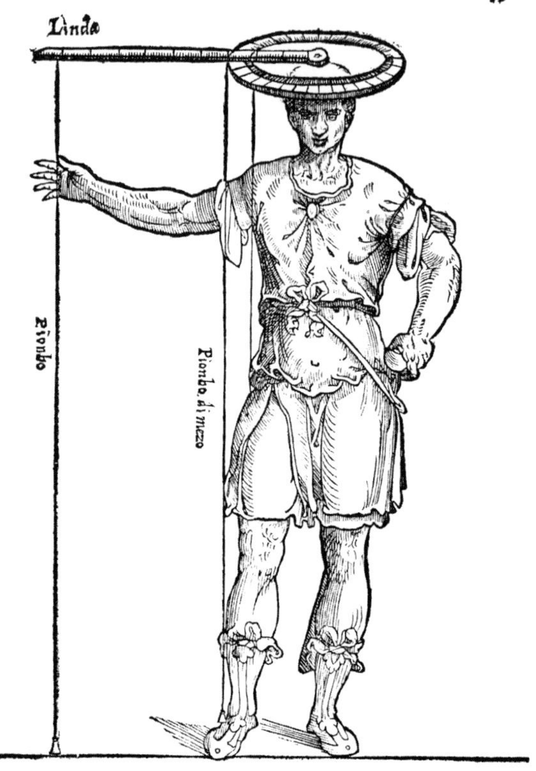

3.5
Alberti's measuring device for statuary replication, from the first publication in print of Alberti's *De Statua*, translated into Italian and illustrated by Cosimo Bartoli in *Opuscoli morali di Leon Batista Alberti gentil'huomo firentino* ..., (Venice: Francesco Franceschi, 1568), 299.

sculptors should take down the spatial coordinates of as many points of a model as needed, then use this numeric log alone to make copies. Alberti claims that by using his technology, different workshops in distant places can be tasked to reproduce parts of the same statue, which, when assembled, would perfectly fit together.[39] The idea to use a digital scan to notate and replicate full-round objects in space must have seemed outlandish, or worse, when Alberti first conceived of it, and until recently scholars and historians were quick to dismiss Alberti's treatise on sculpture as an abstruse exercise on the proportions of the human body. Indeed, the scanning process Alberti describes would have been laborious to carry out in his time, and Alberti does not even try to explain how the resulting numeric file could have been used to actually execute a sculpture. That would have been a most arduous task using traditional artisanal tools, either by subtraction (*per forza di levare*) or addition of material (*per via di porre*).[40] Several equally unfruitful attempts to develop a similar replicating technology are recorded at the onset of the mechanical age. Samuel F. B. Morse, a Yale College graduate and noted painter before he rose to fame for his contributions to telegraphy, strived to develop a marble carving machine that would automatically and exactly replicate "perfect copies of any model," and he even tried to patent one in 1823, together with a New Haven craftsman (only to find out that the patent would have infringed on another one).[41] Morse was appointed professor of painting and sculpture at New York University on October 2, 1832; his title was changed to professor of literature of the arts of design in 1835.[42] In one of his lectures on the arts (delivered in 1826) he mentions Leonardo's *Treatise on Painting*, first published in 1651 together with Alberti's *De Statua*.[43] Several circular saw mills and programmable lathes were developed throughout the nineteenth century, but on March 28, 1842, Henry Dexter, of Boston, appears to have patented a fairly precise replica of

Alberti's scanning machine, suitably doubled in that instance by a similar replicating apparatus—with no better success than his Florentine predecessor.[44] For Alberti's technology would have required a seamless connection between a number-based scan and a similarly number-based fabrication process, which no manual tool or mechanical technology could effectively provide. Today's cheap and increasingly ubiquitous 3-D scanners and 3-D printers work exactly that way.

3.5 The Digital Renaissance of the Third Dimension

In the early 1990s architects of the first digital age started tweaking existing engineering (and, often, medical) 3-D scanning technologies to adapt them to the scale and processes of then nascent computer-aided design and prototyping. The first experiments in Frank Gehry's office to scan and digitize Gehry's own hand-made sculptural maquettes are still legendary,[45] but digital manufacturing tools in the 1990s were bulky and expensive; the subtractive CNC milling machine, which many avant-garde architects then championed, can only cut and carve matter out of a solid block, and its performance is limited by the number of axes on which its drill can operate. The real breakthrough in the history of digital stereopoiesis came only in the twenty-first century, with a new generation of 3-D scanners and additive fabrication tools. The best-known manufacturer of desktop 3-D printers, MakerBot, was founded in 2009. Its first preassembled 3-D printer, the MakerBot Replicator, cost $1,749 when it was launched, in January 2012,[46] but when Michael Hansmeyer and Benjamin Dillenburger wanted to 3-D print the twenty-square-meter room of their first *Grotto Prototype* (2012–13: see chapter 2), they had to resort to a much bigger, industrial grade Voxeljet machine, normally used to 3-D print disposable molds for the mass production of metal components.[47] Many now see 3-D printing as a turning point in the history of technology—as

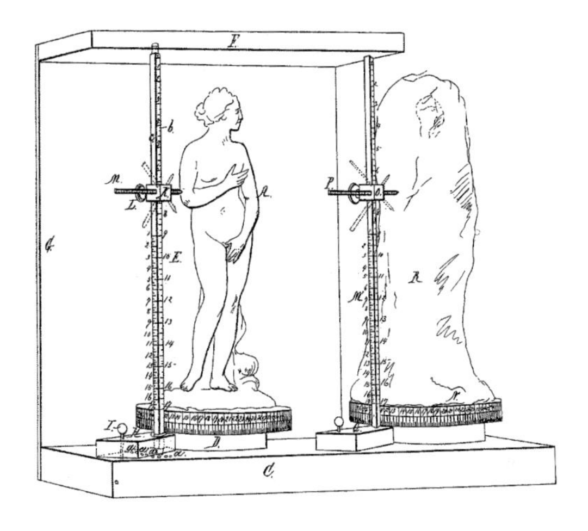

3.6
US Patent 2,519, “Apparatus for Sculptors to be Employed in Copying Busts, etc.,” to Henry Dexter of Boston, March 28, 1842. Source: United States Patent and Trademark Office, http://www.uspto.gov.

US president Barack Obama said in his 2013 State of the Union address, a new technology that “has the potential to revolutionize the way we make almost everything.”[48] At the same time, however, the seamless connection between 3-D scanning, 3-D printing, and virtual reality tools is also likely to change the way we *see* almost everything, and represent and know the world around us.

The first irruption of affordable 3-D scanning in mainstream architectural education came, once again, from the creative appropriation of a machine originally developed for other purposes. Launched in 2010, the Microsoft Kinect was a mass-marketed, motion-sensing device sold as a plug-in for videogame consoles; it allowed players to interact with a videogame through motion and voice.[49] It contained an ingenious depth sensor that worked by triangulations, using a laser projector and a camera. The measurements it provided proved very accurate at short distances, and several applications were soon developed to use the machine as a stand-alone scanner. In 2012 students in schools of architecture around the world were using the Kinect to scan small objects of all sorts; the scans were imported into standard CAD software, and edited at will.[50] Depth-sensing technologies are evolving quickly; some use traditional laser or infrared beams, and calculate distances based on the time of rebound (also known as “time of flight”), or, increasingly, by reading the difference in phase between outgoing and returning beams; some use triangulations, like the earlier Kinect machines did, with a laser beam sending a marker to the target and a camera, in another vertex of the triangle, to read it (variants of this method are known as structured-light depth sensors); and so on. All these methods require some specific hardware to send or read signals; some can measure actual distances, while others can render solid shapes proportionally but cannot calculate measurements. New technologies are also being introduced to reconstruct three-dimensional volumes from a stream of simple photographs—that is, from a sequence of planar, perspectival projections.

As Wheatstone had already explained, our mind builds up solid volumes by interpreting two synchronic perspectival images taken from two slightly different vantage points, but in the absence of the second image, similar results can be obtained by comparing monocular images taken in a quick sequence from a moving vantage point.[51] Contemporary computational depth sensors can reenact both processes, in the case of the latter by using any existing digital camera as the only input tool. Snapshots taken from a camera suitably moving around the same object can be collated and merged, and its solid shape (but not its actual dimensions) reconstructed by computational triangulations. This seems to be the mathematical logic behind Autodesk's 123D Catch, a software that can generate a point cloud of any stationary, full-round object from snapshots taken from a camera moving around it.[52] The data, fed by a cell phone, can in turn be imported into proprietary CAD software, edited and solid modeled using wireframes, meshes, or subdivisions, 3-D printed, or, with additional measurements and some manual work, converted into scaled architectural blueprints in plans, elevations, and sections—thus fulfilling Pietro Accolti's dream of 1625. Project Tango, by Google, uses a more complicated, purpose-built hardware (including both motion sensing and depth sensing) to obtain a 3-D model of interior spaces from the natural movements of someone walking through it.[53] Using similar technologies, the Madrid-based company Factum Arte has specialized in museum work and conservation facsimiles;[54] the ScanLab at the Bartlett School of Architecture in London, known for its 3-D replicas of landscapes and large-scale artifacts, is also developing a range of visual tools to exploit all sorts of images derived from solid models.[55]

Three-dimensional models can be visualized using perspectival images or axonometric or parallel projections, navigable and scalable at will; other interfaces may include virtual reality or augmented reality tools. But each of these planar renderings is only, in a sense, a way to *visit* an actual model—each visit being different, based on contingent requirements and intentions. The only stable part, and the keystone and kernel of the whole system, is the 3-D model itself. And, of course, any 3-D model can be 3-D printed, in full or in part, and (in theory) at any scale.

The modern technologies we were familiar with until recently would typically allow us to take a snapshot of any full-round model—say, a cat—and print that out as a photographic, perspectival picture. Soon, most cell phones will take 3-D scans, not pictures; and keeping, editing, and sending a statuary selfie will cost the same as saving, editing, viewing, sharing, or even printing a pictorial one. Today's technology already allows us to take a snapshot scan of a cat and print that out, right away, as a sculpture;[56] soon we shall visualize a full-round model of said cat in three-dimensional, VR simulations. Planar images still have many practical advantages over 3-D models: for example, a picture of a cat printed on photographic paper is lighter than a statue of the same printed in resin, plastic, or sandstone, at cat-size—or even smaller. Likewise, just like most traditional (planar) photographs stopped being printed long ago, to be shown and seen only on electronic displays, a sequence of electronic images is often the easiest and least cumbersome way to simulate or navigate a 3-D model.

The first commercial light field camera, the Lytro, was released in 2012; it was advertised as a camera that allows users to refocus every picture, and marginally shift the vantage point of each picture, *after* the snapshot has been taken.[57] It was not a commercial success, partly due to the limits of light-field

3.7
ScanLAB Projects, London, and BBC, *Rome's Invisible City—3-D Scan of the Mithras Temple Hidden Below Modern Rome* (2014).

3.8

ScanLab Projects, London, *3-D Scan of Greenpeace's icebreaker* The Arctic Sunrise (2011).

technologies for depth sensing, but the spirit of the game was clear: when you take a picture that way, you do not project it onto a screen (the Albertian way) once and for all; you create a 3-D model in space that you can eventually visit at will—looking in different directions and moving around it (in Albertian terms, rotating the central ray and changing the point of view). At the time of writing (summer 2016) some sports events have already been broadcast live in virtual reality (including from the Rio Olympics), to be experienced through head-mounted displays. The degree of immersivity supported by these VR technologies is variable: the vantage point of the end user may be fixed or movable and the angle of rotation of the head more or less wide; the headsets do not have to be stereoscopic, although it helps if they are.

Alongside virtual reality, a new generation of head-mounted displays supports augmented reality and mixed reality reenactements:[58] the ways to exploit and experience a 3-D model, once it is made, are countless, and as long as we have eyes to see, we shall keep using monocular images (better if paired and synced for stereoscopy) for all kinds of reasons and tasks. But the competitive edge that projected images enjoyed for centuries over 3-D models was as much due to physical lightness as to data lightness: from Alberti until recently, projected images were the easiest way to capture, record, transmit, and replicate all sorts of full-round originals, because projections (perspectival or other) compress a lot of spatial information into small and portable planar files—most of the time, as small as a piece of paper. That still holds true, but it matters less and less, because data is now so easy to gather and so cheap to keep and copy.

Almost one generation ago, the rise of digital photography first dented the cultural and technical primacy of modern projected images. Digital photographs may look like film-based (argentic) photographs, but they are no longer the indexical

3.9
PHOTOMATON® S.A.S. corporate website page detailing the French company's 3-D photo booth service (2015). PHOTOMATON® is a registered trademark owned by the Photomaton Company. Courtesy of the Photomaton Company.

trace left by a beam of light on a photosensitive surface; they are, regardless of the way they are captured, the occasional and always ephemeral end product of a number-based algorithm. As the late William J. Mitchell pointed out at the very beginning of the digital turn, verisimilitude, or the indexical value of proof, which was the main source of the power of modern photography, was simply obliterated by the technical logic of digital photography.[59] Cultural critics and modernist art historians for the most part stubbornly declined to take notice, and thus fell out of kilter with the new world of digitally generated images that we have been living with for the last twenty years. Much more of that is happening now, and it may henceforth be more difficult—even for modernists—to look the other way. For after losing their indexical value during the first digital turn, digital images are now losing all their residual notational functions.

At the end of the Middle Ages the conflation of a new technology for capturing and compressing images, and of a new technology for reproducing them, changed the history of the West. Today, the conflation of new technologies for capturing and reproducing reality directly in three dimensions, without the mediation of images, is likely to have similar epoch-making consequences. Around the mid-sixteenth century Jacopo Pontormo, the lunatic Florentine painter, could claim that while God needed three dimensions to create nature, painters needed only two to recreate it: which, he concluded, "is truly a miraculous, divine artifice."[60] As we can represent and reproduce the world just as it was made—in three dimensions—we need far less of that artifice today. Three-dimensional models have replaced text and images as our tools of choice for the notation and replication, representation and quantification of the physical world around us: born verbal, then gone visual, knowledge can now be recorded and transmitted in a new spatial format.

4 THE PARTICIPATORY TURN THAT NEVER WAS

Digital mass customization may well have been one the most important architectural inventions of all times: originally intended to change the way we manufacture physical objects of daily use (teapots, tables, buildings), the technical logic and the culture of mass customization have already changed—or at least subverted, upended, and disrupted—almost every aspect of the world in which we live. Regardless of these epoch-making consequences, however, the most conspicuous avatar of the first digital turn in architecture and design was the creation of a new architectural style—the new digital style of smooth and curving, "spliny" lines and surfaces. This style, now called parametricism,[1] continues to this day, with ever-increasing degrees of technical mastery and prowess: ideas and forms that twenty years ago were championed by a handful of digital pioneers today engender architectural masterpieces at a gigantic, almost planetary scale. Yet, inherent in the very same technical definition of digital mass customization is a crucial authorial conundrum, which from the start has accompanied and challenged the rise of the parametric worldview. Any parametric notation contains by definition an infinite number of variations (one for each value of a given parameter). Who is going to design them all? Who is going to choose the best among so many options? Early in the new millennium, with the rise of the participatory spirit of the so-called Web 2.0,[2] many thought that collaboration and interactivity could be the answer: the customer, client, or any other stakeholder would be called on to intervene and participate in the design process, to "customize," or co-design, the end product within the framework and limits set by the designers or administrators of the system.[3] At the time of writing (2016), it already seems safe to conclude that this transition from mass customization to mass collaboration didn't happen in architecture and design. But while the design professions mostly rejected this option, digitally driven mass customization has taken over the world.

4.1 The New Digital Science of the Many

Francis Galton was a late-Victorian polymath, half-cousin of Charles Darwin, a descendant of Quakers, and, among other things, one of the fathers of eugenics, the inventor of the weather map, and of the scientific classification of fingerprinting. In one of his last writings he also tried to account for a statistical quirk he had stumbled upon: in some cases, it seems possible to infer an unknown but verifiable quantity simply by asking the largest possible number of random observers to guess it, and then calculating the average of their answers. In the case studied by Galton, a group of farmers tried to estimate the weight of an ox put up for auction at a country fair, and the arithmetical mean of all answers came closer to the actual weight than each individual guess. In modern statistical terms, the accuracy of the average increases in proportion to the number of opinions expressed, regardless of the expertise of, or the specific information available to, any of the observers.

Galton's experiment suggests that, if there is a way to gather the knowledge of many, a group may end up knowing more than even the most knowledgeable of its members; and if this is true for the weight of an ox displayed at a country fair, this may also apply to more complex cases, including queries with no known answers. Possibly due to his eugenic creed (and the related belief, still shared by many, that sometimes the opinion of some should count more than the majority's), Galton failed to remark that his theory of collective intelligence validates the most radical interpretations of the principle of universal suffrage: the more people vote, and the more votes are counted, the better the decisions a democratic community can make, even on subjects about which most voters know nothing at all. In the case of elections, the technology needed to tap the wisdom of crowds is an easy one, and relatively time-tested: each person should simply cast a vote (by a show of hands in a public assembly or, in absentia, by ballot, mail, etc.).

The classical theory of the marketplace offers another case in point, which Galton similarly failed to take into account: market prices are determined by the fluctuations of supply and demand, and they are more reliable when more market-makers can more easily and freely interact with one another. In Adam Smith's classic formulation, this interaction ("the invisible hand of the market") transforms the individual interests of each into the best decisions for all (in this instance, the price that will best allocate limited resources in a free market economy). To help buyers and sellers meet, the world of commerce has always been quick to adopt all kinds of new information and communication technologies, but the spirit of the game has not changed significantly since the beginning of time: traders reflect and express all the information they have in the prices they agree to, and the crying out of bids on an exchange (or any other technologically mediated version of the same) is the market's classic tool to garner the knowledge of many—from as many sources as possible.[4]

For the last twenty years or so, computers and the Internet have offered unprecedented possibilities for gathering and exploiting the wisdom of crowds, and—as vividly recounted in a recent best-seller by the *New Yorker* economist James Surowiecki—in more recent times Galton's curious experiment has become a common reference for adepts of the so-called Web 2.0, or the participatory web.[5] Before the economic meltdown of the fall of 2008, many also thought that the newly deregulated financial markets, enhanced by digital technologies, had attained an ideal state of almost perfect efficiency, where the frequency and speed of limitless and frictionless transactions would make market valuations more reliable than ever before.[6] Later events have proven that those new digital markets were no more infallible than all previous ones, but in other instances digital technologies may have made a better use of Galton's theory. The success of Google's search engine is famously due less to its power of information retrieval—any computer can do that—than to the way Google ranks its findings. These are now increasingly personalized (adapted to the customer's profile, web history, or location, for example), but the original PageRank algorithm[7] prioritized Google's search results based on the sheer quantity and quality of links between Internet (HTML) pages. As in the scholarly system of footnote cross-referencing, which is said to have inspired the PageRank algorithm, these links were first established by the authors themselves, then added to by many others; hence Google's celebrated claim to use "the collective intelligence of the Web to determine a page's importance."[8]

Taken to the limit, the apparently banal combination of search and ranking has staggering epistemic implications, which have been discussed in the preceding chapters: in a world where all events are recorded and retrievable, the search for an exact precedent may better predict future events than an analytic calculation of consequences deducted from general causal laws, rules, or formulas. Indeed, in many cases the search for a social precedent (rather than for a material one, as seen in chapter 2, section 2.4) has already replaced the traditional reliance on the rules or laws of a discipline: for example, when we choose a linguistic expression or syntagm based on the number of its Google hits, we trust the wisdom of crowds instead of the rules of grammar and syntax. Of course, the rules of grammar and syntax themselves are born out of the authority of precedent, as for the most part they formalize and generalize the regularities embedded in the collective or literary use of a language—a process that in the case of living languages unfolds over time and continues forever (the most notable exception being the invention ex nihilo of the rules of classical Latin in the Renaissance). But today a simple Google search on an incommensurably vast corpus of textual sources can effectively short-circuit the laborious scientific process of the constitution of the rules of a language, thus making all traditional sciences of language unnecessary. Not by science, but by search we can draw on the collective intelligence of a group, be apprised of the frequency of an event (in this instance, a linguistic occurrence within a community of speakers), and act accordingly.

4.2 The Style of Many Hands

The making of objects appears to be, intuitively, a quintessentially participatory endeavor, because large or complex objects in particular must often be made by many people, and their design must draw from the skills of many specialists. Yet contrary to this matter of fact, starting with early modern humanism, Western culture has built a cultural system where works of the intellect, regardless of their material complexity, are expected to be ideated by an individual author and to be the expression of just one mind.[9] This applies to media objects (texts, images, music, etc., which today exist primarily as digital files) as well as to physical objects (chairs, buildings, clothing, cookies, or soft drinks), as since the rise of the modern authorial paradigm in the Renaissance it is expected that physical objects should be designed prior to being made; all their creative value being thus ascribed to their design, recipe, or formula, which is a pure piece of information—a media object like any other.[10] But the ways to solicit, collect, and marshal the opinions of many in the making of intangible informational objects, such as drawings or scripts, is not made any easier by the immateriality of the process. The collective creation of a piece of intellectual work cannot be reduced to the expression of a vote or a number to be counted or averaged, respectively—even though many of today's "social media" are in fact reduced to doing just that. This is where many thought that today's digital technologies could be a game changer.

Unlike documents in print, digital notations can change anytime, and every reader of a digital file can, technically, write on it or rewrite it at will: in the digital domain every consumer can be a producer.[11] Moreover, unlike radio or television, the Internet is a symmetrical information technology—whoever can download a file from the Internet can, theoretically, upload a similar one. This technical state of permanent interactive variability offers unlimited possibilities for aggregating the judgment of many, as theoretically anyone can edit and add to any digital object at will. But if interventions are to be open to all, and open-ended in time, as per Galton's model, any change may randomly introduce faults or glitches (the equivalent of a statistical deviation), which will in turn be edited out only by some subsequent intervention. The development of any such participatory object will hence inevitably be erratic and discontinuous. Fluctuations will diminish as the number of interventions accrue, correcting one another, and the object converges toward its state of completion (the analogue of a statistical mean), which will be achieved at infinity, when all participants have interacted with all others in all orders, and the knowledge of all has ideally merged into a single design.

As infinity is seldom a viable proposition for human endeavors, most models of aggregatory versioning must introduce a cut-off line somewhere and maintain some form of moderation or supervision all along. Yet, no matter the varying degrees of authority that may be exerted to curb its randomness, the logic of convergence to the mean of the statistical model still defines most practical strategies deriving from it. In order to self-correct, the process must remain open to as many agents as possible and for as long as possible. This is easy for some media objects, such as a *Wikipedia* entry or open-source software, which may remain variable forever: so, just as in Adam Smith's model of the marketplace, in the case of *Wikipedia* the "invisible hand" of an infinite number of minute, frictionless interactions will convert the ambition and limits and occasional folly of many individual writers into a globally reliable, collective, and anonymous source. But that open-ended model can hardly apply to design notations, for example, as their development must necessarily stop before building can start. It also follows that at each point in time any of such collaborative objects will be to some extent deviant from its theoretical state of completion, hence, to some extent, faulty. Any digital object in a state of permanent development is, by definition, never finished or ever stable, and will be forever functioning only in part—hence it is destined to be, in some unpredictable way, always in part nonfunctioning.

Indeed, daily experience confirms that most digital objects seem to be in a state of permanent drift. Beta versions, or trial versions, are the rule of the digital world, not the exception—even commercial software, which after testing is marketed and sold, is constantly updated, and paying customers are often referred to users' forums (that is, to the wisdom of crowds) to address malfunctions in proprietary, commercial software. This is not accidental: most things digital are permanently evolving into new versions that may not be in any way more stable or permanent than the earlier one, as the open-ended logic of "aggregatory" design must by definition apply at every step and at all times. It may appear counterintuitive that the iron law of digital notations (at its basis a binary system that consists only of zeroes and ones, and nothing in between) should have spawned a new generation of technical systems deliberately meant to work by hit-or-miss, and a technical logic based on an intrinsic and quintessential state of ricketiness. Yet, against the dangers of unpredictable mutations, contemporary digital culture has already largely, albeit tacitly, integrated the good old precautionary principle of redundancy: most digital systems offer an often confusing variety of ways to achieve the same result—evidently hoping that glitches may not affect all at the same time.[12] As most things today are digitally made, designed, or controlled, this evolutionary, "aggregatory" logic of permanent interactive versioning may well be one of the most pervasive technical paradigms of our age, of which traits can already be detected in every aspect of our technical environment—and, increasingly, in our social practices as well.

There was a time, not long ago, when every household in Europe and North America had a fixed telephone line. Barring major calamities, that telephone line was expected to work—and even in a major calamity, it was the last technology to fail. For some years now, synchronous voice communication has been delivered by a panoply of different means: analog fixed lines, where still extant; fixed lines that look like telephones but are in fact VoIP in disguise; GSM cell phones, 3G or above "smart" phones, DSL or cable-connected computers, wirelessly connected computers (via Wi-Fi, local broadband, WiMax, etc.), each using a different network and protocol to communicate. We need so many options because some of them, when called upon, will most certainly not work. Not surprisingly, in the digital domain, versioning and redundancy are far from being simple technical strategies. Given the technical conditions described above, they are fast becoming a mental and cultural attitude, almost a way of thinking.

The logic of the "aggregatory" mode of composition posits that each agent be allowed to edit at will, but anecdotal evidence, corroborated by a significant body of texts and software created by interactive accrual, suggests that each new edit may more easily add new data than erase the old, except to the extent that the erasure is functionally indispensable. As a result, obsolete data are often carried over and neglected, rather than deleted; in software writing whole chunks of script are simply shunted, but left in place, so they may eventually be retrieved. Some authors seem to be applying the same principle, more or less deliberately, to texts—and not only for electronic publication. In the humanistic and modern authorial tradition most objects of design (whether physical or media objects) bear the imprint of an intelligence that has organized them with rigor and economy, and the precise fitting of their parts is often the most visible sign of an invisible logic at work, followed by the author who conceived the object and nurtured its development. According to a famous essay by Alexandre Koyré, first published in 1948, precision is the hallmark of modernity in all aspects of life and science.[13] But collaborative design environments seem to be following a different logic. Particularly in the digital model of open-ended aggregation, the effectiveness of the result is achieved not by dint of authorial precision, but through approximation, redundancy, and endless participatory revisions.

Objects that are made by many hands show the signs of many styles. Many *Wikipedia* entries, for example, are extraordinarily informative, and most open-source software works well, or no worse than any equivalent commercial software. But regardless of their use value, the textual writing of most *Wikipedia* entries, as well as the mathematical writing of most open-source software, is redundant, circuitous, and fragmentary. In contemporary software parlance, a quick fix that will debug or update a piece of script is commonly called a "patch," and patchiness is fast becoming a common attribute of all that is composed digitally: patches of script are added when and where needed, and so long as the resulting text does what it was meant to do, no one cares to rub off the edges or to smooth out its texture any further. Sometimes, as in the case of software scripting, the patchiness of the script will be seen only by specialists; sometimes the end product itself may appear patched up, as most *Wikipedia* entries do. The typical *Wikipedia* entry is written by many people and by no one in particular, but by each one individually.[14]

4.3 Building: Digital Agencies and Their Styles

This new, "aggregatory" way of digital making may equally affect buildings, insofar as digital technologies are implied in their design, realization, or both.[15] Indeed, digitally designed architecture is even more prone to participatory modes of agency, as from its very beginning the theory of digital parametricism has posited a distinction in principle between the design of some general features of an object and of some of its ancillary, variable aspects: digital mass customization (as defined by Gilles Deleuze's and Bernard Cache's theory of the *objectile*) implies a model of layered authorship, or "split agency," where the primary author designs a generic (parametric) object, and one or more secondary authors, or interactors, adjust and adapt some variable aspects of the original notation at will. This authorial model has accompanied digital design theory from its beginnings in the early 1990s and has prompted several forms of hybrid agency, some more open to collaboration, some less so.[16]

Participatory agencies are particularly prominent in a family of software known as BIM, or Building Information Modeling, which for the most part was developed independently of the more tectonically oriented CAD/CAM of the 1990s. The spirit of BIM posits that all technical agents participating in design and construction should collaborate using a shareable information model throughout all stages of a project, and that design decisions should be agreed upon by all parties (clients, designers, and contractors).[17] In such instances, architectural authorship might take the form of some consensual "leadership," curiously resembling the organization of labor that prevailed in late-medieval building sites before the rise of Alberti's modern authorial paradigm.[18] Architects frequently blame BIM for its bureaucratic bias, and the technology was indeed developed primarily for managerial, not design, purposes. The participatory logic of BIM also differs from the digital "aggregatory" model in some essential aspects. Participation in BIM-based design is by invitation only, and invited participants are limited to technical agents—even though one could imagine a variety of other parties interested in the development of a building, including end users, communities, and even citizens. Moreover, contrary to the principle of aleatoric accrual from independent contributors in the digital, open-ended model, BIM decision making is based on a negotiated consensus among few parties, a practice that may remind some of the more traditional modes of "design by committee," which few architects ever cherished—even the less Ayn Randian of them.

According to a proverb still popular among the design professions, "A camel is a horse designed by a committee." Apparently implying that camels do not look good and that committees are not good designers, the proverb is also curiously attributed to Sir Alec Issigonis, the idiosyncratic engineer who conceived one of the most remarkable camels in the history of the automobile, the 1959 Mini. In fact, on closer scrutiny, it seems that committees are more likely to endorse consensual, generic solutions than a specific, and possibly brilliant, but unconventional idea, such as a camel or the Mini (the former strange looking if compared to a horse, the latter if compared to British cars of 1959, yet both more suitable than their standard mainstream counterparts for traveling in the desert or facing fuel shortages in London after the Suez Canal crisis, respectively). BIM design processes, as currently envisioned, encourage technical agents to come to some form of middle-ground compromise at all stages of design and invite expert feedback and social collaboration, but for the same reason they may also favor team leadership over competence, or safe and bland solutions to the detriment of riskier but innovative ones—including unusual, bold, or assertive formal solutions, shapes, and styles. This regression toward a consensual mean may be the most significant contribution of BIM technologies to architectural visuality—a leveling effect that may apply to the making of the most iconic, monumental buildings (Gehry Technologies has developed its own BIM platform, which it also sells to third parties) as well as to countless nondescript utilitarian buildings, where BIM technologies are already employed by the construction industry—without fanfare but with significant time and cost savings.

The digital model of open participation by accrual does not work that way, and in architecture it might lead to very different visual consequences. Following the time-tested example of open-source software, architectural authorship might be succeeded by some form of mentoring or supervision, where one agency or person initiates the design process, then monitors, prods and curbs, and occasionally censors the interventions of others. The social, technical, and theoretical implications of this way of design by curatorship are vast and mostly unheeded.[19] Its effects on designed objects, and on our designed environment at large, could be equally momentous. Regardless of the amount of supervision that must be brought to bear at some point, either by human or by machinic intervention, physical objects designed by participatory aggregation will most likely show some signs of the approximation, redundancy, patchiness, and disjointedness that are the hallmark of all that is designed by many hands. And there are reasons to assume that most designers today would not like that—much as they may not be happy to relinquish the legacy of the authorial privileges the design professions have so laboriously struggled to acquire over time. After all, it took centuries, starting with Renaissance humanism, to establish modern architecture as an authorial, allographic, notational art of design, which medieval and classical architecture never were; and which non-Western architecture—that is, architecture outside of the influence of Western humanism—is often not to this day. In the course of the twentieth century architects were remarkably successful in adapting the Albertian, humanistic idea of design to the industrial mode of production, and they managed to impose that authorial, notational way of making as the dominant model of the architectural profession throughout the world. Many of today's designers may just not be inclined to give that up without a fight.

Of course, disjunctions and fragmentation are not unknown in contemporary design. Aggregation was a distinctive stylistic trait of architectural deconstructivism. Deconstructivist theories in architecture, particularly through the well-known Derrida-Deleuze connection, were not only the precedent but also the direct cause, and almost the midwife of the digital revolution in architecture as we know it. Traces of the fractured disjunctions of deconstructivism can be found in much of the digitally inspired architecture that followed in the 1990s and beyond—in works of Zaha Hadid, Wolf Prix, Frank Gehry, or Peter Eisenman himself. Yet this purely visual analogy, or even the historical continuity it portends, may be misleading. The paratactical disjointedness of Eisenman's deconstructivist formalism of the late 1980s was quintessentially authorial. It followed from and was based on the excruciatingly detailed and finicky execution of the design of one mind, in the purest Albertian notational tradition.[20] Deconstructivists may aim at designing and notating complexity or even at representing or interpreting indeterminacy; but few designers trained in the Western tradition would aim at or even consider letting complexity just happen. Indeed, most of the design strategies discussed in this book are, indeed, *design* strategies—strategies for staying in charge.

With different nuances, and with some exceptions—of which adaptive fabrication and hard-core morphogenetic design may be the most conspicuous—the second digital turn in architecture is largely about finding new ways to design, not about finding ways to not design. Not designing may be an ideological statement or a profession of faith, but it is unlikely to ever become a well-paid profession; while some digitally intelligent designers pride themselves on using open-source software, few or none author open-ended design—architectural notations that others could modify at will.[21] As will be seen in chapter 5, the ideas of permanent variability, parametric mass customization, and digitally driven mass collaboration that designers test-drove during the age of the first digital turn are now spreading in all areas of contemporary society, economy, and politics. Designers do not need to follow that trend because they started it, and they already know what it is about. As it happens, right now, they are busy inventing something else.

5 ECONOMIES WITHOUT SCALE: TOWARD A NONSTANDARD SOCIETY

A few years ago I held a visiting professorship at a major US university located in one of those remote, bucolic retreats where Europeans would love to go on holiday for a few days in August and Americans still oblige their daughters and sons to spend some years of their youth. My weekly commute included a short hop by plane from a big airport to the airstrip serving the rural region around that campus, and as I wanted to travel with a small but rigid carry-on suitcase, I checked in advance which plane was used on that route. I found out the logistic was one I am very much familiar with: the flight would be on a CRJ 200, a most uncomfortable regional jet derived from an older executive plane, stretched and retrofitted with close-packed rows of four seats. The overhead bins in that plane are of course similarly minuscule, yet compliant with the sub-IATA standard of 55 by 40 by 20 centimeters for cabin baggage, which appears to have been tacitly adopted by many European low-cost carriers, and which until recently was also enforced by Lufthansa and some of its partner airlines in Europe. Not surprisingly, this is also the size of my German-made Rimowa suitcase, which Rimowa sells, incorrectly, as an IATA-sized carry-on. To make a long story short, I knew for certain that my luggage would fit in the overhead locker of the CRJ 200's cabin, and said as much when I was asked to check it at the gate of the plane.

The flight attendant, a young woman, was not forthcoming. "You will check it like everyone else," she barked. "The flight is almost empty," I offered, "and my luggage is made to measure for the overhead bins of your plane: it will fit just fine and get in no one's way." But the flight attendant was evidently having a bad day. She screamed an ill-tempered challenge: "Come on in, try it! It will not fit; it cannot fit, because it has wheels." But it did fit, perfectly and effortlessly, and the locker snapped shut with a crisp, audible click. Still, she was not convinced. "You cheated," she raged. "I would not know how to cheat in such an instance," I think I answered. Then, half to myself, as the flight attendant abandoned the scene, I mumbled that the reason the suitcase fitted so well was that both the maker of the plane and the maker of my luggage had been following the same standard. But this excess of information, along with my European accent and the keyword "standard," used in that mildly confrontational context, was an unfortunate combination. It also drew the attention of another passenger, who was sitting in the first row; an elderly gentleman wearing a button-down Oxford shirt *and* a tie (albeit not a bow tie) with some sort of blue blazer in lieu of the tweed jacket more usually favored by academics on the campus I was traveling to. He looked up from his *Wall Street Journal*, eyed me with some interest, and asked, puzzled, "Are you a socialist?"

I don't remember precisely what I answered, but it was something bland and noncommittal—after all, I thought, the man might be a donor to the college I was working for, or the father of one of my students. But the work I had brought on board stayed in the aforementioned overhead bin during the flight, while I pondered the apparent paradox so poignantly emphasized by the comment of that untweedy traveler—a man who looked like he was used to being listened to. In the story I know, standardization is something inherent in the technical logic of the Industrial Revolution, and it has no particular political bias. All countries in the course of their industrial development have embraced and favored technical standards: in the twentieth century, the assembly line, for example, was adopted by socialist and capitalist economies alike. How, then, had my fellow traveler come to resent standardization, a staple of modernity, and why did he perceive it as a collectivist ploy?

5.1 Mass Production, Economies of Scale, Standardization

The reason why industrial standards are so central to modern mass production is well-known. Most industrial technologies use mechanical matrices (stamps, molds, casts, etc.) to reproduce identical copies. Matrices have a cost, and once made, it makes sense to keep using them to amortize this cost by spreading it over as many copies as possible. Economies of scale increase in proportion to the number of copies, hence in this mode of production it is cheaper to make more identical copies of fewer different items. One means to this end is the creation of catalogs of interchangeable, combinable items that can be sold to a larger market: for example, standard nuts and bolts, or standard steel I-beams, which can be used across many different industries, or—indeed—standardized carry-on bags that can be used on all planes, instead of having to make and sell a different size of bag for each airline or airplane cabin. Such industrial standards require agreements and regulations, which can be arrived at in many different ways.

More contentious is the argument that a similar logic also applies outside the realm of mechanical, matrix-based technologies of mass production. Do human bodies, for example, really perform better when they are obliged to repeat the same gesture many times over? Frederick Taylor's *Principles of Scientific Management* (1911) assumed they do: in spite of the evident fact that human bodies do not function like machines, assembly lines force workers to take their cue from the mechanical tools of mass production they use, apparently not without some gains in efficiency. I once knew a dentist who claimed she was the best in the world at pulling out tooth number 16—and not its symmetrical counterpart 26—because since the beginning of her career she had been doing only that, and her hand had become perfect and infallible in that gesture. More contentious still, however, is the application of the technical logic of mechanical standardization to domains that are neither technical nor mechanical, yet appear to have been, to some extent, standardized in the course of the last two centuries in order to achieve similar economies of scale. Pricing, as both a social practice and a cultural technology, is a case in point.

At the beginning of time, pricing (or barter) was the result of a free, unconstrained negotiation, or dialogue, conducted in person by two parties in the physical presence of the goods they would exchange. As dialogue is an art and haggling is an often theatrical figure of dialogue, agreement between two hagglers famously depends on the talents of each actor, with the more talented being rewarded with a better deal. But dialogue, dealing, and haggling take time; and if each transaction takes forever, few transactions will ever take place. This limitation, coupled with a recognition that some social groups may profit from some degree of predictability as to the results of such dealings, explains why, over time, societies have invented endless ways to limit the freedom of interpersonal economic exchanges and to regulate some prices. For example, in late medieval and early modern Europe the prices of many goods and services were strictly regulated by urban guilds. The power that pre-industrial guilds wielded over economic life was so pervasive that at some point (notably around 1776, when Adam Smith published his *Wealth of Nations*) many concluded that making and trading should be deregulated, and individuals should be allowed to exchange goods and services as they wished. According to the economic doctrine of classical liberalism, as it was formulated around that time, individuals freely pursuing their own self-interest also maximize the common good, and free prices are the best way to match supply and demand and to allocate limited resources. In more recent times neoliberal, neoconservative, and neoclassical economists have used similar arguments against state regulation and (until 1989) against socialist, centralized planning.

However, modern free pricing has also long been fighting a losing battle against other enemies from within, right at the core of free-market economies. Unrelated to cartels and monopolies—an easy target long singled out by communist propaganda—the most successful opponent of free pricing in modern free-market economies has been, oddly, the department store. Like many large corporations, department stores freely compete with one another externally, but impose strict centralization and almost socialist planning and regulations internally, within the ambit of their own organization. Dealings among branches of the same company may not be visible to outsiders, but the retail price at which the department store decides to sell an item to the general public is one of its most conspicuous, distinguishing traits. Since the late nineteenth century this has been a fixed price—nonnegotiable, centrally determined, and the same for all. Salespeople in a department store are not allowed to bargain, and customers cannot bid a different price. In compensation, department stores allow them to walk in and out at will.

5.2 The Rise and Fall of Standard Prices

The fixed-price revolution is often attributed to Aristide Boucicaut's Bon Marché department store in Paris, which still exists, though its original name is today somewhat misleading.[1] The names of other French department store chains (Monoprix, Prisunic) bear witness to this retail revolution, which in fact evolved over the course of the nineteenth century in all industrialized countries as a consequence of the growing size of retail outlets. An independent merchant who owns the stuff he buys can easily decide the price of the stuff he sells. But as stores expanded and had to train more and more salespeople to deal with customers, it proved more efficient to diminish their bargaining leeway and in the end to give them none, centralizing pricing decisions. Selling at fixed prices evidently saves time, allowing for more sales; and many transactions at a less-than-perfect, standard price may ultimately yield more profit than fewer sales at a better price.[2] Moreover, many transactions in a department store are so small that they do not warrant the time and cost of bargaining. Fixed prices also seem more honest, transparent, and democratic—as no one, not even the foolish and the gullible, pays more than others—and they are often seen as symbols of a certain idea of modernity, whether that be a cause or a consequence of their adoption. Be that as it may, statistics indicate that in contemporary industrialized economies around half of all consumer expenditures for goods and services is transacted at fixed prices, which further suggests that the number of individual, fixed-price transactions far exceeds that of individually negotiated ones, today limited to a few big-ticket items such as cars and real estate.[3] Economists also take it for granted that most customers in developed countries prefer the ease and speed of "modern" fixed prices to the aleatoric opacity of "traditional," bazaar-like open-ended negotiations. At least this was the case until recently.

Because once again, as architects have learned in the daily practice of computer-based design and fabrication, digital postmodernity is upending the game, creating a new environment and new practices that are much closer to our premodern, artisanal past than to our modern, industrial present. The principles of nonstandard seriality in digital making are now almost twenty years old, and their potentials and limits have been tested by the digital avant-garde in architecture since the early 1990s. Unlike mechanical making, digital making is rarely matrix-based, hence using file-to-factory digital technologies it is theoretically possible to mass-produce variations, within limits, at no extra cost. As we know, digital file-to-factory technologies (popularized today by the diffusion of cheap, distributed 3-D printing) offer no economies of scale: the unit cost of the first item in a series is the same as the unit cost of all subsequent ones, whether they be all identical or all different.[4] However, it now appears that this technical logic may also apply to some social practices where no material production is involved, and with similarly disruptive consequences.

Pricing decisions used to be time-consuming, labor-intensive, and costly activities. But this is no longer the case if dealing and pricing can be automated and the final price is agreed on by humans dealing with computers—or by computers dealing with other computers. Algorithmically based, digitally networked transactions cost very little—in fact, almost nothing; and they are increasingly sophisticated and precise, as computers can now cull and process more data in a few seconds than an expert salesman could in a lifetime. Prices can be algorithmically determined wherever and whenever a web-enabled device (at the time of writing, something as small as a smartphone) can connect to the Internet—which is the same as saying, almost everywhere and at all times. Evidently, this means that even good old fixed prices can now be permanently updated and adapted in real time to fluctuations in supply and demand, and synchronized on a global scale without the glitches and quirks arising from local specificities or lack of communication. The more data these frictionless, digital markets are based on, the more efficient they tend to become, and in theory they would become perfect if all market makers could participate in the same market at the same time.[5] An even better version of seamlessly unfettered pricing would be the permanent, global auctioning off of everything that can be sold. The Praetorian Guard once put the Roman Empire up for auction, but today eBay could offer it to a much larger pool of possible bidders, increasing the statistical chances of selling it to a better candidate than the hapless Milanese tycoon Didius Julianus, who bought it in the year 193 and came to no good.[6]

An additional step in computer-based automated pricing could simply adopt the same technology that Google or Amazon already use to customize their search results and book recommendations, respectively. While Google famously claimed—until recently—to use the "collective intelligence" of the web to rank findings, search results are now increasingly prioritized based on location, search history, or preferences expressed by the searcher.[7] And Amazon's suggestions for further reading based on each customer's past purchases are often more pertinent than the advice the most expert bookseller or even a close colleague or friend could offer.

When I was a teenager I lived in a town of 50,000 inhabitants, with one high school and one good bookshop. I profited vastly from the expertise and insight of that one bookseller—which was actually a team of two, a husband and wife. They were learned, cultivated people and they probably knew something about books, but they also knew me, my father, mother, grandparents, all my high school teachers, and most of my friends. The advice I received in that shop was always valuable, and in a few memorable cases crucial. After the book I should read was agreed on, so was the price, which was never the listed price—students often got discounts, and some more than others. I see no reason why Amazon would not do the same, at some point. Perhaps it does already—even though recent experiments of automatically or algorithmically customized prices in other markets have been met with ferocious opposition.[8] Yet in my opinion, many customers like me are getting used to the notion that, when buying online, fixed prices have already ceased to exist—air travel being the most conspicuous example of this, and iTunes the most notable exception. Sure, the logic whereby most online prices change all the time is unfathomable—but so was that of the small-town bookseller I remember so well. And the expertise and intuition of the astute haggler who knows who can and will pay more for a given item may already be matched by "machine learning" and by a computer-based analysis of a customer's buying patterns. The data is there, and its computational interpretation, or "mining," costs little.

5.3 The Digital Mass-Customization of Social Practices

As the cost of each ad hoc, computer-based pricing decision and of each automated transaction is almost equal to zero, computer-based buying and selling, just like computer-based manufacturing, is indifferent to economies of scale. On the Internet, the transaction cost of buying or selling almost nothing from or to a single person, or in bulk, or from or to many at the same time, is almost the same and equally negligible in all cases. So, for example, it is possible today to rent a car for just sixty minutes, which would have been unthinkable until recently; theoretically, this means that renting ten different cars for one hour each should not cost more than renting the same car for ten consecutive hours, and in fact, through Zipcar this is almost so. Likewise, renting one room for the night from a lessor who manages a stock of just one room is no more expensive than the cost per night of renting a room for one month from a lessor who manages a stock of thousands—in fact, through Airbnb, it often costs less, albeit probably not for the reasons under discussion here. Remarkably, before the rise of zero-cost Internet-based transactions, such instances of micro-leases could only have been conceived as personal favors among friends ("Could I borrow your car till 4 p.m.?" "Yes, you may have my apartment this weekend."). This is because no serious business would have managed individual transactions that are worth so much less than the cost of manually processing the administrative overhead applying to each.

Once again, it would appear that digital tools may help us to recreate some degree of the organic, spontaneous adaptivity that allowed traditional societies to function, albeit messily by our standards, before the rise of modern task specialization. Preindustrial societies did not need hotel chains, not only because fewer people traveled, but also because the few that did were mostly taken in, somehow, by the local population: aristocrats by next of kin, pilgrims by convents, artisans by fellow guild members, and, when necessary, the army could be billeted with all of the above. Today, many similarly messy, micromanaged arrangements could become viable again using the Internet, Big Data, and algorithmic pricing: freeways could bill ad hoc not only for the use of faster traffic lanes, but also for minute driving decisions, such as passing or resting at a service area, and traffic rules could change all the time, locally or due to contingencies, so long as each driver, or car, is in the electronic loop all the time; cab drivers could sell their services in real time using a permanent auction system based on location and time, thus matching supply and demand more efficiently than the present archaic, theatrical, and capricious American practice of whistling, screaming, and haggling on the street, which, by the way is forbidden in most countries in continental Europe;[9] bus drivers (and soon driverless buses) could adapt their schedule and itinerary as needed, so long as demand is algorithmically optimized and all users are apprised of changing travel plans—which is what the driver of the bus that served the Alpine village where I went on holiday as a child always did, without any technology other than common sense, and without regard for a myriad of state and local regulations. At its limit, one could also imagine a system where trains do not leave and stop on schedule, but follow computationally negotiated, ad hoc surveys of all concerned, letting users decide in real time when the train should leave and where it should stop. And so on.

Economists have not failed to recognize that the Internet and Big Data are changing retail by allowing more and more customized pricing aimed at more and more fragmented markets—which is something akin to what postmodern philosophers, who did not have retail in mind, and whom economists seldom read anyway, called "the fragmentation of master narratives."[10] But the trend I have been describing goes far beyond the often unwieldy multiplication of customer choices already largely exploited by late-modern marketing tools and strategies. To understand what is at stake, one should look at the bigger picture. Once again, the combination of connectivity and digital computation favors the simulation, and at times almost the reenactment, of ancestral, pre-industrial, and premodern ways of making and dealing. Computer-based design and fabrication eliminate the need to standardize and mass-produce physical objects, and in many cases this has already created a new culture of distributed making that some call digital craftsmanship. Likewise, computer-based commerce eliminates any need to aggregate supply, demand, and items of trade, and this favors personal, one-to-one, zero-cost transactions that are in many ways similar to the original and ancestral dialogue between two persons who freely agree to exchange some goods or services under conditions of their choosing. In this model, every transaction is a nonstandard one-off, a special case: there are no standards and there need not be, simply because there would be no gains if the transaction were scaled up to include more people or more items. Of course, not every transaction needs to be negotiated each time anew, just as not every mass-produced physical article needs to be mass-customized. It is simply that transactions and physical items that would profit from being made to measure now can be, and at no extra cost. Standardization is no longer a money-saver, and variations are no longer a money-waster. In fact, in a digital techno-cultural environment the opposite is often the case.

And to go back to the incident that sparked these thoughts: even if I had a 3-D printer at home, I do not think I would like to melt and remake the rigid polycarbonate case of my carry-on every time anew based on the plane I would be flying in. I would prefer to have, say, three ready-made suitcases for three global standards of cabin bins. The standardizing spirit of the mechanical age would then live on for many items of hard and fast mechanical mass production. Yet, in the bigger scheme of things, that untweedy *Wall Street Journal* reader might have had a point after all. If some Uber-like technology can already wipe out all taxi cartels and taxi and limousine commissions, it is not a long shot to imagine that sooner or later some Bitcoin-like technology may do the same with national currencies and all related monetary authorities. One might generalize and conclude that many regulating functions that governments have taken upon themselves over time were necessary in the past to achieve economies of scale, but those regulations, and their regulators, are now becoming equally unnecessary in a digital environment where economies of scale no longer apply. If that is so, today's neoconservatives and libertarians could hire computers to fire some of the government bodies they abhor. In fact, they may not even need to do so; and they may not be the only interested party. As the threshold of aggregation that apparently maximized economies of scale during the Industrial Revolution was that of the nation-state, the modern nation-state may be the first redundancy in a post-standard, post-scale techno-social environment. City-states, which thrived in late medieval

Europe as free economic areas outside of feudal and imperial control, were stifled and phased out by the rise of modern centralized states. Back then, city-states were too small to regulate and standardize mechanical mass production—they did not have the size to compete on scale. Yet today the few still extant, and a few newly created outside of Europe, all appear to be doing well.[11] We'll see.

POSTFACE: 2016

Technical change does not happen in a vacuum. Inventions may be random accidents, but a new technology can only take root and thrive if many need it and use it. In this sense, as the anthropologist André Leroi-Gourhan argued a long time ago, every technology is a social construction: innovation only occurs when technical supply matches cultural demand, and when a new technology and new social practices are congruent within the same techno-social feedback loop.[1] The first digital turn started almost one generation ago, and some crucial social changes can already be ascribed to it—never mind in which order of causality. An embryonic outline of the new social, political, and economic environment related to today's technologies of digital mass-customization has been suggested passim in this book, and more specifically in chapter 5.

As I finished writing this book, I witnessed global changes of a magnitude that my generation had never experienced. Too young to remember the Cuban missile crisis, I grew up in a stable, bipolar world in which European and American safety and prosperity were guaranteed by the threat of Mutually Assured Destruction (MAD) and by a number of social covenants, both national and international. The fall of the Berlin Wall may have been an earthquake, but seen from the West its severity was mitigated by self-satisfaction: the Cold War was over because we had won it. Now it's our turn to be on the losing side. Just as in the 1990s the fall of the Soviet Union and of the Warsaw Pact brought about havoc, misery, and war in parts of Eastern Europe, today Western Europe is heading for the abolition of the European Union and of NATO—likely, with similar consequences. At the time of this writing, the United Kingdom is envisaging the forcible deportation of millions of legal European residents; on the other side of the Atlantic, the US has just elected an abrasive, unpredictable billionaire without any political experience to the most powerful office in the world. Do the technologies I describe in this book—in their pristine, inchoate form, mere technical tools we use to design and make better and cheaper physical objects—have something to do with any of this?

They do. And I am not even thinking here of the influence of today's new communication technologies. Evidently, there is a link between the format of the new media and the kind of content they may carry (in the case of Twitter, the medium of choice of the new US president, discourse being limited to statements of 140 characters). And while modernist consensus was based on monodirectional, one-to-many mechanical or electromechanical media (like print, or Hertzian radio and television), where the same message was sent to all at the same time, postmodernist dissensus is based on bidirectional, many-to-many, asynchronous digital media (like today's so-called social media), where there is no technical difference between the statement of a scientist and that of a quack. Since the 1980s, the vulgarians' version of postmodernism has advocated the disappearance of factual evidence—as a noted scholar I know recently said during a seminar at the university where I teach, "Facts are always filtered and mirrored by the never ending multiplicities of our rhizomatic, gendered, and eroticized identities, subjectivities, and individualities" (verbatim). So there you go: If this is what we want, this is what we get—instantly taken up, multiplied, and spread far and wide by social media as by a superhuman war machine.[2] But this is not my point.

As discussed at length throughout this book, digital mass-customization allows us to mass-produce variations at no extra cost (that is, at fixed marginal costs), thus eliminating economies of scale from a digital design and production workflow. More recently, the same logic has been applied to all kinds of financial operations and social practices, ideally leading to an almost zero-cost transactional environment, where the upfront cost of setting up and processing a transaction is so small that it need not be amortized by size (see chapter 5). These two technological changes together sound the death knell for the modern market-based industrial system, and with it, of modern liberalism, of modern capitalism, and many related things, ideas, and tenets of modernity. Extrapolating from these premises, some in recent times have been nurturing the notion that a market economy based on competition and pricing may soon be replaced by a social economy based on collaboration and commonality: a zero-cost economy where everyone is happy to donate labor and knowledge because almost everything is free.[3] Against all odds, a participatory, zero marginal cost economy is already within reach for many informational goods (that is, goods that are pure data), as well as for some renewable energies: a small, locally managed hydroelectric power station can produce electricity almost for free and almost forever after its initial cost is paid back, if someone can take care of it on a daily basis. But the zero marginal cost society is unlikely to ever become an enticing prospect for the design professions, for two different reasons.

First, land and building materials are not renewable. Iron and concrete cannot be pulled from thin air, the way electricity can, and, no matter how much we recycle, raw materials will always be in limited supply, hence expensive. Second, all professions have a vested interest in self-preservation, so design professionals tend to think that they cannot be replaced by a machine, and that their own expertise has a unique value and a cost. This is where digital technologies have already started to prove them wrong. The adoption of the first, still primitive forms of artificial intelligence for simple technical processes and tasks (of the kinds that have been discussed here) is not only subverting our ways of making. It is also, surreptitiously, pushing us toward a new way of thinking: it is training us to *think in a different way*, following a new, post-scientific logic. As discussed in chapter 2, the rift between these two logics is such that, until and unless we learn to cope with this new science, we are likely to make a terrible mess of it, and of what remains of our own traditional scientific logic at the same time. This is where the epistemic dislocation brought about by the rise of artificial intelligence (AI) in daily life resonates with today's political and social turmoil. In all aspects of science, technology, and life, we are behaving like that old friend of my grandfather's who, after riding a bike to his fields all his life, sold his farm, bought a sports car, and died by crashing it on day one. Alfa Romeos of the 1960s were temperamental, and it took time to learn to drive one. It may take even longer to learn to drive AI, instead of being driven by it—as I fear we are doing now.

The first demotion of traditional, humanistic authorship in most arts and sciences, as discussed in chapter 4, comes from crowdsourcing, and its efficacy is largely proven: the authorial *Encyclopaedia Britannica*, where each entry is authored and signed by a respected specialist, is no longer in print; *Wikipedia*, where most entries are anonymously compiled and edited by whoever has the time and will to do so, is thriving. But on closer scrutiny, the wisdom of *Wikipedia* is not really a wisdom of crowds: most entries are de facto curated by fairly traditional scholar communities, and these communities can contribute their expertise for free only because their work has already been paid for by others—often by universities or other public institutions. In this sense, *Wikipedia* is only piggybacking on someone else's research investments (but multiplying their outreach, which is one reason for its success).

The new kind of science that is inherent in most of today's AI applications, and at the core of what I call the second digital turn, is a different matter altogether. In our traditional way of thinking, facts are laboriously collated, sifted, compared, and selected, then generalized and formalized: the apex of this sorting process is a theory, often compressed into mathematical formulas, which we use to predict similar events when similarly describable. Computers don't do that; they search for a precedent. We, in turn, use computation to simulate as many fictional precedents as needed when no actual one is on record, and when we do not have the time to compare some of these results ourselves, we ask computers to randomly test as many as possible, ad infinitum if necessary, knowing that at some point we shall find one or two that will more or less do what we need. And computers can do all that—simulations and optimization included—so fast, that this dumb procedure is often the best, and in fact, for some classes of problems, the only one available.

Traditional artisans, who were not engineers and did not use math, proceeded by trial and error: you make a chair, and if it breaks you make it again, and again, until you make one that does not break. Structural calculations are smarter and cheaper because you design a chair before you make it, thus knowing in advance that when you make it, the chair won't break. This is why, over time, we came to trust engineers more than artisans. But today's computational tools work like artisans, not like engineers. Increasingly, so do we: we are relearning the serendipitous charms of heuristics. Plenty of training in digitally empowered architectural studios today extols the magical virtue of computational trial and error. Making is a matter of feeling, not thinking: *just do it*. Does it break? Try again … and again … and again. Or even better, let the computer try them all (optimize). The idiotic stupor (literally) and ecstatic silence that are often the primary pedagogical tools in many of today's advanced computational studios rightly apprehend the incantatory appeal of the whole process: whether something works, or not, no one can or cares to tell why.

As discussed in chapter 2, this abdication of traditional causal reasoning works just fine in plenty of cases. Computational simulation and optimization (today often enacted via even more sophisticated devices, like cellular automata or agent-based systems) are powerful, effective, and perfectly functional tools. Predicated as they are on the inner workings and logic of today's computation, which they exploit in full, they allow us to expand the ambit of the physical stuff we make in many new and exciting ways. But while computers do not need theories, we do. We should not try to imitate the iterative methods of the computational tools we use because we can never hope to replicate their speed. Hence the strategy I advocated in this book: each to its trade; let's keep for us what we do best.

Instead, tragically, the opposite seems to be happening. In all aspects of contemporary culture, and most remarkably in economics and politics, theories today are universally reviled. With theories, all the makers and markers of theory, and many ingredients of theory-making, are being equally and drastically demoted: facts, observation, verification, demonstration, proof, experts, expertise, experience, competence, science, scholarship, mediation, argument, political representation, and so on—in no particular order. Why waste time to argue? Ask the crowds. Why waste time on a theory? Just try it and see if it works. But computational simulations are made of bits and bytes, and can be rerun at will; the next atomic blast in physical reality may not allow for a retrial.

London, December 29, 2016

NOTES

CHAPTER 2

I have discussed some of the ideas presented in the introduction and in chapter 2 in one or more of the following articles: "Digital Phenomenologies: From Postmodern Indeterminacy to Big Data and Computation," *GAM* 10 (2014): 100–112; "Breaking the Curve. Big Data and Digital Design," *Artforum* 52, 6 (2014): 168–173; "Big Data and the End of History," *Perspecta* 48, *Amnesia* (2015): 46–60; "The Digital is Our Stuff," in *Fluid Totality*, ed. Zaha Hadid and Patrik Schumacher (Basel: Birkhäuser, 2015): 20–25; "The New Science of Form-Searching," *AD* 237 (2015): 22–27; "Christian Kerez's Art of the Incidental," *Arch+* 51 (2016): 70–76; "Excessive Resolution," *AD* 244 (2016): 80–83; "The New Science of Making," in *LabStudio: Design Research Between Architecture & Biology*, ed. Jenny E. Sabin and Peter Lloyd Jones (London and New York: Routledge, 2017).

1. Luca Pacioli, *Summa di Arithmetica Geometria Proportioni & Proportionalita* (Venice: Paganino de Paganini, 1494), fol. 33. Pacioli may not have been the first to describe this method in print: see David Eugene Smith, *History of Mathematics*, vol. 2: Special Topics of Elementary Mathematics (New York: Dover, 1958), 141 (first published Boston: Ginn and Co., 1925).
2. The arithmetic of al-Khowârizmî (c. 825) was first translated into Latin as *Liber Algorismi*: see Smith, *History of Mathematics*, 2: 9–11 and 78–88. The meaning of algorism in early-modern mathematics is unrelated to the modern meaning of the term "algorithm," which is defined, for example, in Merriam-Webster's Collegiate Dictionary (11th ed., 2014) as "a procedure for solving a mathematical problem" and also, more broadly, as "a step-by-step procedure for solving a problem or accomplishing some end esp. by a computer."

3. The invention of decimal fractions in the West is generally attributed to Simon Stevin, and commas or periods started to be used to notate decimal numbers slightly later: Stevin, *De Thiende* … (Leyden: C. Plantin, 1585), also published in French by Plantin the same year as a part of Stevin's *Arithmetic* (*L'Arithmétique de Simon Stevin, … Aussi l'Algèbre … Ensemble les quatre premiers livres d'algèbre de Diophante d'Alexandrie, maintenant premièrement traduicts en François. Encore un livre particulier de la Pratique d'arithmétique, contenant entre autres, les tables d'Interest, la Disme et un Traicté des incommensurables grandeurs; avec l'explication du dixiesme Livre d'Euclide* [Leyden: C. Plantin, 1585]). The first description of logarithms is in John Napier, *Mirifici Logarithmorum Canonis Descriptio; ejusque usus, in utraque trigonometria, ut etiam in omni logistica mathematica … explicatio* (Edinburgh: Andrew Hart, 1614), and the first logarithmic tables in print were published in Henry Briggs, *Arithmetica Logarithmica, sive Logarithmorum chiliades triginta, pro numeris naturali serie crescentibus ab vnitate ad 20,000: et a 90,000 ad 100,000 … Hos numeros primus invenit … Iohannes Neperus Baro Merchistonij: eos autem ex eiusdem sententia mutavit, eorumque ortum et vsum illustravit H. Briggius* (London: William Jones, 1624). See Dirk J. Struik, *A Concise History of Mathematics* (New York: Dover, 1987), 89 (first published New York: Dover, 1948), with further bibliography.
4. Pierre-Simon de Laplace, *Exposition du système du monde* (Paris: Imprimerie du Cercle-Social, IV–VI [1796–98]), vol. II:5, IV:266. ("[Kepler …] eut dans ses dernières années, l'avantage de voir naître et de profiter de la découverte des logarythmes, artifice admirable, dû à Neper, baron écossais; et qui, réduisant à quelques heures, le travail de plusieurs mois, double, si l'on peut ansi dire, la vie des astronomes, et leur épargne les erreurs et les dégoûts inséparables des longs calculs; invention d'autant plus satisfaisante pour l'esprit humain, qu'il l'a tirée en entier, de son propre fonds. Dans les arts, l'homme emploie les matériaux et les forces de la nature, pour accroître sa puissance; mais ici, tout est son ouvrage." During his last years, Kepler could see the birth and profit from the discovery of logarithms, an admirable artifice due to Napier, a Scottish laird. By reducing to a few hours the labor of many months, logarithms double, so to speak, the life of astronomers, and spare them the errors and discomfort that are inseparable from long

calculations. This invention is even more pleasant to the human mind, if we consider that it is entirely of an intellectual origin. In the arts, man increases his powers by dint of materials and forces he derives from nature; but in this case, all comes from him alone. [My translation]).

5. Elizabeth Eisenstein similarly argued that the increasing precision of astronomical measurements in the seventeenth century was due to the circulation of ponderous astronomical tables in print, rather than via scribal copies, starting with Kepler's *Rudolphine Tables* (Ulm, 1627). Elizabeth L. Eisenstein, *The Printing Revolution in Early Modern Europe* (Cambridge: Cambridge University Press, 1983), 217.
6. The invention of modern "binary arithmetic" is attributed to Leibniz, and to an essay Leibniz published in 1703 in particular: "Explication de l'Arithmétique Binaire, qui se sert des seuls caractères O et I; avec des Remarques sur son utilité, & sur ce qu'elle donne le sens des anciennes figures Chinoises de Fohy" in *Histoire de L'Académie Royale des Sciences, Année MDCCIII* (Paris: Jean Boudot, 1705), 85–89. However, not much was made of the binary notation before it started to be used for relay- and switch-based electric computing as of 1937–40.
7. Here, and elsewhere in this book, the expression "cultural technology" refers to instruments that consist exclusively of formalized procedures, or to similarly formalized processes interacting with some external devices (examples: the alphabet, logarithms, folding a shirt, or Alpine skiing). The notion of *Kulturtechnik* was popularized in European media studies by the work of Friedrich Kittler (1943–2011) and of his school.
8. On the origin and different meanings of the expression, see Victor-Mayer Schönberger and Kenneth Cukier, *Big Data: A Revolution That Will Transform How We Live, Work, and Think* (Boston: Houghton Mifflin Harcourt, 2013), 6 and footnotes.
9. See Walter J. Ong, *Orality and Literacy: The Technologizing of the Word* (London: Methuen, 1982; 2nd ed., London: Routledge, 1995), esp. chap. 4, "Writing is a technology," 81–83. Citations are to the second edition.
10. For a somewhat similar argument see Robert Logan, *The Alphabet Effect: A Media Ecology Understanding of the Making of Western Civilization* (Cresskill, NJ: Hampton Press, 2004), 218–219: "The articulation of the alphabetic

paradigm of the repeatability of fragmented identical items and the letters in the hardware realm of mechanics created the printing press, assembly line, mass production, and general organizing principle of the industrial age."

11. Writing in 1982, Ong predicted that soon the People's Republic of China would have imposed the use of the Roman alphabet to transcribe phonetically the Mandarin dialect, so as to avoid the production of typewriters using over 40,000 characters (*Orality and Literacy*, 87). Digital technologies have already made this "great leap forward" utterly unnecessary.

12. The theory that the adoption of print with moveable type was one reason for the technological and scientific advancement of Protestant Europe, and of its competitive advantage against other, non-Protestant countries, can be traced to Karl Marx ("Economic Manuscript of 1861–63: Division of Labour and Mechanical Workshop. Tool and Machinery," in Karl Marx and Frederick Engels, *Collected Works*, vol. 33 [London: Lawrence and Wishart, 1991], xix–1169, 402–404), and is today a historiographical commonplace. But the necessary technical premise to the early modern development of print with moveable type would in turn have been the phonetic alphabet, which according to Walter Ong "was invented only once ... by ancient Semites and perfected by Ancient Greeks" (*Orality and Literacy*, 91). On the shift from print to digital media in contemporary Arabic script, see Gráinne Hebeler, "The Alphabet Is Dead ... Orality and Graphic Representation in the Middle East," unpublished master's thesis, Bartlett School of Architecture, 2015.

13. The main (and, to date, only) sources for the history of Gmail are the Gmail official blog and the "History of Gmail" entry in *Wikipedia*. See *The Official Gmail Blog*; "Welcome to the Official Gmail Blog," blog entry by Bill Kee, July 3, 2007, accessed October 17, 2015, http://gmailblog.blogspot.in/2007/06/welcome-to-official-gmail-blog.html; "History of Gmail," *Wikipedia*, last modified August 19, 2016, https://en.wikipedia.org/wiki/History_of_Gmail.

14. Gmail offers a "Delete forever" command (October 17, 2015, https://support.google.com/mail/answer/7401?hl=en), but there is some controversy on how and when the deletion from the company's servers is carried out. See discussion, accessed October 17, 2015, http://www.zdnet.com/article/does-delete-forever-in-gmail-really-mean-it, and a number of forums and online discussions on the topic.

15. The *Categories* is one of Aristotle's six books on logic: Aristotle, *The Organon, 1: The Categories on Interpretation; Prior Analytics*, ed. and trans. Harold P. Cooke and Hugh Tredennick (London: William Heinemann; Cambridge, MA: Harvard University Press, 1938; Loeb Classical Library 325), 1–109.
16. Classical and medieval logic did not admit propositions with singular subjects (i.e., statements where the subject is an individual, rather than a class), hence in classical or medieval terms, the individual cannot be the object of science. According to Joseph M. Bochenski, a syllogistics based on singular premises became widespread only in the seventeenth century (starting with Port Royal)—although some isolated precedents are known (Bochenski, *Formale Logik* [Freiburg: K. Alber, 1956]; *A History of Formal Logic*, ed. and trans. Ivo Thomas [Notre Dame, IN: University of Notre Dame Press, 1961], 232).
17. See Frances Yates, *The Art of Memory* (London: Routledge and Kegan Paul, 1966), 173–199.
18. "Although this seems close to the view of a madman, such a view is nevertheless a real component of Ramist mentality, always implied and always operative." Walter J. Ong, *Ramus: Method and the Decay of Dialogue* (Cambridge, MA: Harvard University Press, 1958), 192. See Mario Carpo, *Metodo e ordini nella teoria architettonica dei primi moderni* (Geneva: Droz, 1993), 58–63. On early modern arborescent classifications see also Lina Bolzoni, *La stanza della memoria* (Turin: Einaudi, 1995), 26–86.
19. Mario Carpo, *Architecture in the Age of Printing* (Cambridge, MA: MIT Press, 2001), 109–111.
20. Thus the Protestant theologian Matthäus Richter on the use of print to disseminate the words of Martin Luther (Matthäus Richter [Matthaeus Judex, or Iudex], *De Typographiae inventione, et de praelorum legitima inspectione, libellus brevis et utilis* [Copenhagen: Johannes Zimmermann, 1566]).
21. Carpo, *Architecture in the Age of Printing*, 109.
22. Carpo, *Metodo e ordini*, 65–82, with further bibliography; Yates, *The Art of Memory*, 129–160; Lina Bolzoni, *Il teatro della memoria: Studi su Giulio Camillo* (Padova: Liviana, 1984). Camillo once famously tamed and hypnotized a lion he had run across in a garden (oddly, in Paris), while everyone else ran away: the story is recounted by Camillo himself in his *L'idea del theatro* (Florence: Torrentino, 1550), 39.

23. See the timeline of Otlet's multifarious activities (in collaboration with Henri La Fontaine, Nobel Peace Prize in 1913; at one point involving Le Corbusier) on the website of the Mundaneum archives, currently held in Mons, Belgium, accessed October 18, 2015, http://archives.mundaneum.org/en/history.
24. See http://www.nytimes.com/2012/03/13/technology/google-to-announce-venture-with-belgian-museum.html.
25. "L'alta adunque fatica nostra è stata di trovare ordine … capace, bastante, distinto, & che tenga sempre il senso svegliato, & la memoria percossa." Camillo, *L'idea del theatro*, 11.
26. The notion that computers can search for any alphanumerical combination in any corpus of any size simply by scanning it all from start to finish, and stopping whenever a given sequence of 0s and 1s is found, albeit theoretically true, must be nuanced in practice. Google, for example, does not scan the entire World Wide Web anew anytime a new string of letters and numbers is typed into a Google search bar. Google searches are based on copies of the web that Google makes at regular intervals; these copies are construed by "spiders" that crawl the web, following the frequency and relevance of HTML links. The same hypertextual logic is at the basis of the proprietary (and secret) PageRank algorithm that Google uses to rank findings, i.e., to show search hits in a different order and to further customize them when they are served to each user (see chapter 4, note 8). The engineering of Google's information retrieval technology has been described in some detail by Google founders Sergey Brin and Larry Page in a seminal scholarly paper they coauthored in 1998, "The Anatomy of a Large-Scale Hypertextual Web Search Engine," in *Proceedings of the Seventh International Conference on World Wide Web* (Amsterdam: Elsevier), 110–117, and *Computer Networks and ISDN Systems*, 30, 1–7 (April 1998): 107–117. Accessed May 1, 2016, http://research.google.com/pubs/SergeyBrin.html. See also John Battelle, *The Search* (London: Penguin Books, 2005), 67–80. Google's "cached" copy of the web is in fact a repository of compressed pages, where each page is indexed (assigned a doc-ID) and resolved into words; each word is then assigned a word-ID and linked to the pages it came from, as in a traditional name index. From that list of words a program generates a lexicon of what should in fact be called keywords; these keywords (in 1998, numbering 14 million) are the terms

that customers can actually use to search Google's copy of the web. So when a search on a term returns no hits, it simply means that that term is not included in Google's current lexicon. Doc-IDs and word-IDs are stored in separate memory units, called barrels (in 1998, Brin and Page used 64 of them). When a search is started on a term featured in the lexicon, the lexicon points to the barrel where that word-ID is kept, and the word-ID in turn refers to the documents where that word occurs; these results are then further ranked based on various criteria of relevance, as mentioned above. So, in order to reduce the time and cost of data retrieval, Google uses a number of shortcuts and contingent strategies that are not substantially different from those a human would use, construing something similar to an automated name index, where some data are sorted somehow by a machine that puts them in a given container and goes back to fetch them when called upon. Brin and Page, however, were adamant that in their original caches or repositories all documents "are stored one after the other," without any hierarchy, and this repository "requires no other data structure to be used in order to access it"; the document key is a fixed width sequential access mode index, where each document is simply assigned a number or an ID. So it would appear that in order to cope with the size and exponential growth of the early web using the technologies that were available at that time, Google engineers had to resort to some traditional small-data tricks and trades; yet the whole system is ideally inspired by the new logic of linear digital searching. In the end, albeit Google prioritizes and customizes findings, searches are conducted on a flat index derived from an uncategorized database (the "cached" copy of the web at a given time) and, unlike books in a library, names in a dictionary, or food in the aisles of a supermarket, that corpus or repository is not classified in any way we would do it—only the search results, or hits, are, and each time anew based on a secret formula that Google tweaks all the time. Brin and Page's paper is now almost twenty years old, and there is no evidence that the nuts and bolts of Google's information retrieval machines still work that way; to the contrary, there is evidence that, in the case of searches on smaller data bases (a search for a name in Google Books, for example, or in a free Gmail account, capped at 15 GB), the whole repository is simply scanned in full to search for a given string of 0s and 1s. In this, the posthuman logic of digital searches is antithetical to the human logic of sorting: as our searching skills are so limited, and manual searches so labor-intensive and time-consuming, humans must systematically sort data to subdivide each corpus into smaller, more easily searchable sets.

27. See, for example, Chris Baraniuk, "How Algorithms Run Amazon's Warehouses," posted on the BBC website on August 18, 2015: "Products on shelves are not organised by category. Instead, they are placed on shelves as if by random. An HDMI cable lies near to five copies of some Harry Potter sheet music. A brand of baby's bottle is across the aisle from a drain water diverter. But there is method to this apparent madness." (Accessed October 23, 2015, http://www.bbc.com/future/story/20150818-how-algorithms-run-amazons-warehouses). The article does not explain how this "apparent madness" works; nor does the recent literature on "chaotic storage" offer more insight on this matter: accessed February 11, 2106, http://www.wired.com/2013/08/spime-watch-chaotic-storage, http://www.thewire.com/technology/2012/12/inside-method-amazons-beautiful-warehouse-madness/59563, and https://www.dhl-discoverlogistics.com/cms/en/course/tasks_functions/warehouse/assignment.jsp.

28. This intuitive notion has been emphasized, in a very different context, by noted epistemologist and computer scientist Gregory Chaitin: "A scientific theory is a computer program that enables you to compute or explain your experimental data. ... A theory is valuable only to the extent that it compresses a great many bits of data into a much smaller number of bits of theory." Chaitin, "The Halting Probability Omega: Irreducible Complexity in Pure Mathematics," *Milan Journal of Mathematics* 75 (2007): 297–298. Chaitin uses this idea to argue that, while it is always possible to prove that a given theory is shorter (hence better) than another, it is never possible to prove that any theory is the shortest (hence the best) of all possible theories. This, according to Chaitin, proves that mathematics has an infinite complexity, and it is not reducible to formal axiomatics; thus vindicating Gödel's incompleteness theorem of 1931. See also Chaitin, "The Limits of Reason," *Scientific American* 294, no. 3 (March 2006): 74–81. Chaitin's arguments have been further interpreted in cognitive terms, thus implying that all human comprehension depends on some form of data compression: see Phil Maguire, Oisin

Mulhall, Rebecca Maguire, and Jessica Taylor, "Compressionism: A Theory of the Mind Based on Data Compression," in *Proceedings of the EuroAsian-Pacific Joint Conference on Cognitive Science*, Turin, Italy, September 25–27, 2015, accessed February 11, 2016, http://ceur-ws.org/Vol-1419/paper0045.pdf. Some of the arguments being discussed here, albeit derived from, and merely pertaining to, the history of media and of cultural technologies, may lead to similar conclusions. Remarkably, Chaitin does not seem to have noticed that data compression, which was as central to the history of science as it appears to be central to the performance of our cognitive processes, is now being made obsolete by big data computation. This is one reason why the methodological legacy of so many cultural technologies for data compression is fast disappearing from today's computational science—as well as possibly, at some point, from human cognition. From a quite different vantage point, a brief but striking passage in Gilbert Simondon's *Du mode d'existence des objet techniques* equally highlights the contrasts between the "unstructured," flat nature of all artificial (mechanical) information storage and the organic functioning of human memory, which is always animated by experience and based on an "intelligent selection of forms." But Simondon, writing in 1958, describes machines for the recording of analog sound on magnetic tapes and of moving images on argentic films; he does not envisage or anticipate digital information retrieval or data compression tools, and it appears he never further developed the topic. See Gilbert Simondon, *Du mode d'existence des objet techniques* (Paris: Aubier-Montaigne, 1958), 120–124; my translations.

29. Galileo Galilei, *Discorsi e Dimostrazioni Matematiche intorno à due nuove scienze, attinenti alla mecanica e i movimenti locali* (Leiden: Elzevir, 1638), 116–133 (Discourses and Mathematical Demonstrations Concerning the Two New Sciences of Mechanics and Local Motions).

30. Chris Anderson first argued for a new science of Big Data in his groundbreaking article, "The End of Theory," *Wired* 7 (July 2008): 108–109. That issue of *Wired* was titled "The End of Science," even though the other essays in the "Feature" section of the magazine did little to corroborate Anderson's vivid arguments. Anderson's main point was that ubiquitous data collection and randomized data mining would enable researchers to discover

unsuspected correlations between series of events, and to predict future events without any understanding of their causes (hence without any need for scientific theories). A debate followed and Anderson retracted some of his conclusions (Schönberger and Cukier, *Big Data*, 70–72 and footnotes). From an epistemological point of view, however, what was meant by "correlation" in that debate did not differ from the modern notion of causality, other than in the practicalities of the collection of much bigger sets of data, and in today's much faster technologies for data processing. Both classical causation and today's computational "correlation" posit quantitative, cause-to-effect relationships between phenomena; and both the old (manual) methods of scientific inquiry, and today's computational ones, need some hypotheses to select sets of data among which even unexpected correlations may emerge. Evidently, today's computational processes make the testing of any such hypotheses much faster and more effective, but the methodological and qualitative changes that would follow from such faster feedback loops between hypotheses and verification were not part of that discussion. A somewhat similar but more promising debate is now taking place in some branches of applied technologies, such as structural engineering: see section 2.4, esp. notes 34 and 35.

31. The same ontological disclaimer already mentioned on the nature and extension of this entirety applies.

32. See work by Alisa Andrasek, Gilles Retsin, Daniel Widrig, and others in the B-Pro Master's program of the Bartlett School of Architecture, University College London (UCL).

33. Steffen Reichert, Tobias Schwinn, Riccardo LaMagna, Frédéric Waimer, Jan Knippers, and Achim Menges, "Fibrous Structures: An Integrative Approach to Design Computation," *Computer-Aided Design* 52 (2014): 27–39. See also Carpo, "The New Science of Form Searching," in Achim Menges, ed., "Material Synthesis: Fusing the Physical and the Computational," special issue (AD Profile 237), *Architectural Design* 85, no. 5 (2015): 22–27.

34. Finite Element Analysis, or the Method of Finite Elements, works by converting a seamless swathe of homogeneous matter into a lattice or mesh of very small contiguous parts. The equilibrium of this discrete system is then calculated in the usual way, but the high number of parts makes it very difficult to resolve using traditional mathematical tools: although the principles of finite elements date back to the mid-twentieth century, FEA became of practical use only with the advent of electronic computation. FEA is also an early example of digital, "agnostic" science (see note 40), as FEA simulations neither derive from, nor suggest any interpretation of the overall behavior of the structure they describe (other than the equilibrium of the virtual mesh adopted for computational purposes); therefore, FEA simulations do not offer any indication on how any changes in the structure may affect its mechanical performance. In other terms, FEA simulations available today through mainstream software for architectural design will show where a given structure may break, but will not tell why or how to make it better (accessed February 11, 2106, http://www.autodesk.com/solutions/finite-element-analysis, http://www.karamba3d.com). The comparison among the results of many such simulations, however, may suggest ways to improve the design of the structure, either intuitively (in a computer-enhanced, trial-and-error learning process of sorts), or by computational optimization: see note 35.

35. Finding the maxima and minima of a function is the bread and butter of calculus, but problems of shape optimization or structural optimization (for example, finding the shape of the most rigid structure for a given amount of a given material under a given condition of load) involve a huge number of variables, and, except in the simplest cases, solutions to these problems cannot be deducted or calculated from formulas. See François Jouve, "Structural Optimization," *Log* 25 (2012): 41–44. Until recently, engineers would typically tackle these tasks by calculating one initial structure, arbitrarily chosen, which they would then tweak a little and recalculate, and so on, as many times as practically possible until an acceptable solution is found. As structural simulations can now be performed on the screen almost instantaneously, designers today can compare and choose among many more options than would have been possible until recently, effectively trying and testing a virtually unlimited number of possible solutions, at will. Additionally, several methods and computer programs already exist to use computational tools to carry out these trials automatically, rather than manually; hence arriving at better solutions algorithmically, or recursively, rather than intuitively. This is often done by letting the program randomly try out a number of different options (within a given range of variations for each chosen variable), then identifying clusters of values for which that set of variables generates the best results (i.e., the best values in a given field for the function that is being optimized). The program then refocuses on those ranges of variations, dropping (or "killing") all others, and generating new random trials within their ambit; then choosing the best results from within that new range, and so on, ad infinitum. Whenever the program is stopped, the designer can see the best result among all those tried until then, but with no certainty that a better solution may not be found with the next iteration, or with any additional one: this method cannot find the best solution, only one that is as good as it gets; in fact, the designer can choose one solution that he deems "good enough" for his purposes at any time. Such evolutionary methods of optimization are often described in biological, or even Darwinian, terms: see the description of Galapagos, a Grasshopper plug-in, by its author, David Rutten, accessed February 11, 2016, http://www.grasshopper3d.com/group/galapagos. A similar "survival of the fittest," evolutionary method of optimization by random variations is known in web design under the name of A/B testing: see Mario Carpo, "Digital Darwinism," *Log* 26 (2012): 97–105. For evolutionary optimization in structural engineering, see Grant P. Steven and Yi Min Xie, "A Simple Evolutionary Procedure for Structural Optimization," *Computers & Structures* 49, no. 5 (1993): 885–896, and Grant P. Steven and Yi Min Xie, *Evolutionary Structural Optimization* (Heidelberg: Springer, 1997). Evolutionary Structural Optimization (ESO) gradually removes material from a structure as it is being designed. Mark Burry famously used Evolutionary Structural Optimization methods in his work for the continuation of Gaudi's Sagrada Familia in Barcelona: see Mark Burry, Yi Min Xie, et al., "Form Finding for Complex Structures Using Evolutionary Structural Optimization Method," *Design Studies* 26, no. 1 (2005): 55–72. See also the so-called Dynamic Relaxation Method, known for having been used by Norman Foster for the design of the roof of the Great Courtyard of the British Museum in 2000. (I owe some of this information to the unpublished master's thesis of my student Christos Koufidis, Master in Architectural History, 2015, the Bartlett, UCL). The principles of evolutionary algorithms were first studied by John Holland

(*Adaptation in Natural and Artificial Systems: An Introductory Analysis with Applications to Biology, Control, and Artificial Intelligence* [Ann Arbor: University of Michigan Press, 1975]). See also notes 98, 100.

36. See Richard Sennett, *The Craftsman* (New Haven: Yale University Press, 2008).

37. Even the existing methods and software for structural optimization follow essentially the same heuristic, trial-and-error procedure; as they do not deduct results from formalized premises, but simply carry out massive, automated, and open-ended testings of a theoretically unlimited range of randomly variable design hypotheses—with some incremental, or "evolutionary" targeting included in this recursive process: see note 35.

38. The rejection of modern science as a science of universals is central to the postmodern philosophy of Gilles Deleuze and Félix Guattari. In *Milles Plateaux*, in particular, Deleuze and Guattari opposed the "royal science" of modernity, based on discretization ("striated space") to the "smooth space" of "nomad sciences," based on "nonmetric, acentered, rhizomatic multiplicities that occupy space without counting it and can be explored only by legwork," which "seize and determine singularities in the matter, instead of constituting a general form ... they effect individuations through events or haecceities, not through the object as a compound of matter and form." Deleuze and Guattari saw the model of nomad sciences in the artisan lore of medieval master builders, and they had no foreboding of the then nascent new technologies that would inspire digital makers one generation later. See Gilles Deleuze and Félix Guattari, *A Thousand Plateaus*, trans. Brian Massumi (London and New York: Continuum, 2004), chap. 12, "Treatise on Nomadology," 406–409 and 450–451 (first published Minneapolis: University of Minnesota Press, 1987; first published in French as *Milles Plateaux* [Paris: Les Éditions du Minuit, 1980]).

39. Weather forecasting has traditionally included predictive mathematical modeling (known as Numerical Weather Prediction, or NWP) alongside statistical data on local or global weather patterns, unrelated to the mathematical description of the physical phenomena themselves (this latter approach also known as Model Output Statistics, or MOS). In recent times, however, the spike in data input (collected from ubiquitous sensors, crowdsourcing, etc.) and in data processing capabilities has led to expectations of better Numerical Weather Prediction, and to fewer and more timid attempts at using weather data recording for the retrieval of statistically meaningful precedents (as per the heuristic, "search-based" predictive method being discussed here): see for example A. R. Ganguly, E. A. Kodra, A. Agrawal et al., "Toward Enhanced Understanding and Projections of Climate Extremes Using Physics-Guided Data Mining Techniques," *Nonlinear Processes in Geophysics* 21 (2014): 777–795, accessed February 11, 2016, http://www.nonlin-processes-geophys.net/21/777/2014/npg-21-777-2014.html; on the historical modeling of weather patterns evolving in time, E. Kalnay, *Atmospheric Modeling, Data Assimilation and Predictability* (Cambridge: Cambridge University Press, 2002).

40. See D. Napoletani, M. Panza, and D. C. Struppa, "Agnostic Science: Towards a Philosophy of Data Analysis," *Foundations of Science* 16, no. 1 (2011): 1–20; "Artificial Diamonds Are Still Diamonds," *Foundations of Science*, 18, no. 3 (2013): 591–594; "Is Big Data Enough? A Reflection on the Changing Role of Mathematics in Applications," *Notices of the American Mathematical Society* 61, no. 5 (2014): 485–490.

41. As search always starts with, and aims at, one individual event, the science of search is essentially a science of singularities, but the result of each search is always a cluster of many events, which must be compounded, averaged, and aggregated using statistical tools. Thus, there are no lower limits to the level of "precision" of a search (a lower level of precision, i.e., less intension, will generate more hits, i.e., a larger extension in the definition of the set).

42. See note 16. On this aspect of Aristotelian science, see Carlo Diano, *Forma e evento. Principi per una interpretazione del mondo greco* (Venice: Neri Pozza, 1952).

43. See chapter 5, and Mario Carpo, "Micro-Managing Messiness," *AA Files* 67 (2013): 16–19. An earlier version published as "Micro-Managing Messiness: Pricing, and the Costs of a Digital Non-Standard Society," in James Andrachuk, Christos C. Bolos, Avi Forman, and Marcus A. Hooks, eds., "Money," special issue, *Perspecta* 47 (2014): 219–226.

44. See Neri Oxman, "Programming Matter," in Achim Menges, ed., "Material Computation: Higher Integration in Morphogenetic Design," special issue

(AD Profile 216), *Architectural Design* 82, no. 2 (2012): 88–95, on variable property materials; Achim Menges, "Material Resourcefulness: Activating Material Information in Computational Design," in *Architectural Design* 82, no. 2 (2012): 34–43, on nonstandard structural components in natural wood.

45. Menges, "Material Resourcefulness," 42.

46. As Dennis Shelden, chief technology officer of Gehry Technologies, remarked: "The purpose of materials processing—the industrial operations that render trees into 2x4s and ore into metal sheet—is to lower the world's complexity and align its behavior to those geometries for which we have tractable models and numerical solutions." Shelden, "Information Complexity and the Detail," in Mark Garcia, ed., "Future Details of Architecture," special issue (AD Profile 230), *Architectural Design* 84, no. 4 (2014): 94. Shelden continues: "The detailing strategies of today are developed parametrically precisely to package, replicate, and reduce information complexity" (96).

47. This vindicates the premonitions of Ilya Prigogine, another postmodern thinker whose ideas were a powerful source of inspiration for the first generation of digital innovators in the 1990s. See in particular Ilya Prigogine and Isabelle Stengers, *Order Out of Chaos: Man's New Dialogue with Nature* (New York: Bantam Books, 1984); first published in French as *La Nouvelle Alliance: métamorphose de la science* (Paris: Gallimard, 1979).

48. Menges, "The New Cyber-Physical Making: Computational Construction," in Achim Menges, ed., "Material Synthesis: Fusing the Physical and the Computational," special issue (AD Profile 237), *Architectural Design* 85, no. 5 (2015): 28–33. "But what happens if the production machine no longer remains just the obedient executor of predetermined instructions, but begins to have the capacity to sense, react, and act; in other words, to become self-aware?" (29). Menges continues: "Predictive modeling, both as geometric notation and numerical simulation, may eventually be replaced by real-time physical sensing and computational analysis, material monitoring, machine learning, and continual (re)construction" (32).

49. Earlier ideas of mass customization arose in the context of the postmodern pursuit of customer choice, which initially led to a demand for variations compatible with the modes of mechanical mass production then available.

The typical expression of early mass customization was the multiple-choice model, based on a finite number of conspicuous but often purely cosmetic product variations. See Stanley M. Davis, *Future Perfect* (Reading, MA: Addison-Wesley, 1987), where the expression "mass customization" may first have occurred; Joseph B. Pine, *Mass Customization: The New Frontier in Business Competition*, foreword by Stan Davis (Boston: Harvard Business School Press, 1993). At some point in the course of the 1990s it became evident that digital design and production tools (then called CAD/CAM, or "file to factory" technologies) allow for the mass production of endless variations, theoretically at no extra cost. Insofar as it does not employ casts or molds or dies, digital fabrication does not need to amortize the upfront cost of mechanical matrixes, hence in any given cycle of digital fabrication the marginal cost of production is always the same: more identical copies do not make copies cheaper, and variations in the series do not make any new item more expensive. In a sense, digital mass-customization provided a technological answer to a long-standing postmodern quest for product variations; but it is not clear when and how technological supply and cultural demand may have first crossed paths.

The term *mass customization* was first brought to the attention of digital designers by William Mitchell in the late 1990s. See William J. Mitchell, "Antitectonics: The Poetics of Virtuality," in *The Virtual Dimension. Architecture, Representations, and Crash Culture*, ed. John Beckmann (New York: Princeton Architectural Press, 1998), 205–217, in particular "Craft/Cad/Cam," 210–212; William J. Mitchell, *E-topia: Urban Life, Jim—But Not As We Know It* (Cambridge, MA: MIT Press, 1999), see esp. "Mass Customization," 150–152. But both the term and the notion of mass-customization were conspicuously absent from Mitchell's groundbreaking and inspirational *City of Bits* (Cambridge, MA: MIT Press, 1995), which was mostly devoted to the social, spatial, and architectural implications of the migration of activities and functions from physical space to "cyberspace." The idea of mass customization was at the core of the mathematical and technical notion of Deleuze's and Cache's *objectile*, but it does not appear that Bernard Cache, Greg Lynn, or other pioneers of the first digital turn in the 1990s ever adopted the term "mass customization" itself. Starting with his seminal book *Earth Moves*, first published in 1995 (but originally written in 1983), Bernard Cache often used the term "nonstandard" with the same meaning (Cambridge, MA: MIT Press, 1995), esp. 88. Similar notions (but without the use of any of these terms) were also foreshadowed by Lynn in the now famous *AD* 63, "Folding in Architecture" (Greg Lynn, ed., "Folding in Architecture," special issue [AD Profile 102], *Architectural Design* 63, nos. 3–4, [1993], see in particular Lynn's essay on Shoei Yoh, "Shoei Yoh, Odawara Municipal Sports Complex," 94–97, esp. 95). The term "nonstandard" was then popularized by the eponymous exhibition in Paris ("Architectures non standard," Centre Pompidou, Paris, December 10, 2003–March 1, 2004, curated by Frédéric Migayrou and Zeynep Mennan). On the mass production of variable parts at the same cost of standardized, identical ones, see in particular the essays by Migayrou, Lynn, and Cache in the exhibition catalog: *Architectures non standard,* eds. Frédéric Migayrou and Zeynep Mennan (Paris: Éditions du Centre Pompidou, 2003), 28, 90, and 138, respectively. See Mario Carpo, "Ten Years of Folding," introductory essay to the reprint of "Folding in Architecture" (London: Wiley-Academy, 2004), 14–19, esp. 16–18; "Pattern Recognition," in Kurt W. Forster, ed., *Metamorph: Catalogue of the 9th International Biennale d'Architettura, Venice 2004*, 3 vols. (Venice: Marsilio, 2004), 3: 44–58; "The Demise of the Identical: Standardization in the Age of Digital Reproducibility," paper presented at REFRESH. First International Conference on the Histories of Media Art, Science and Technology Banff New Media Institute, September 28–October 1, 2005, accessed February 11, 2016, http://www.mediaarthistory.org/wp-content/uploads/2011/05/Mario_Carpo.pdf, and several times reprinted; "Tempest in a Teapot," *Log* 6 (2005): 99–106; *The Alphabet and the Algorithm*, 93–106; Chris Anderson, *Makers: The New Industrial Revolution* (New York: Random House, 2012), 71–88. On the marketing and psychological implications of multiple choice (without reference to digital technologies), see Barry Schwartz, *The Paradox of Choice* (New York: HarperCollins, 2004). Curiously, today's idea of digital mass customization is prefigured, almost verbatim, in a brief passage at the very end of the first edition of Marshall McLuhan's *Understanding Media* (1964). McLuhan refers to variations in the mass production of automobile tailpipes, which, he claims, are made possible by new computer-controlled "automatic machines." Using such machines, McLuhan claims, "it is possible to make eighty different kinds of tailpipes in succession, as rapidly, as easily, and as cheaply as it is to make eighty of the same kind." (Marshall McLuhan, *Understanding Media: The Extension of Man* [New York: McGraw-Hill, 1964], 314). McLuhan briefly describes the digital manufacturing devices he has in mind, which appear more similar to today's industrial robots than to the numerically controlled milling machines he could have observed back then (see note 81). McLuhan even goes on to note that "the characteristic of electric automation is all in this direction of return to the general-purpose handicraft flexibility that our own hands possess" (ibid.). McLuhan revisited and singularly amplified the same argument in a 1967 article devoted to education and technological change:

> Just as the old mechanical production line pressed physical materials into preset and unvarying molds, so mass education tended to treat students as objects to be shaped, manipulated. … That age has passed. More swiftly than we can realize, we are moving into an era dazzlingly different. Fragmentation, specialization and sameness will be replaced by wholeness, diversity and, above all, a deep involvement. Already, mechanized production lines are yielding to electronically controlled, computerized devices that are quite capable of producing any number of varying things out of the same material. Even today, most U.S. automobiles are, in a sense, custom produced. Figuring all possible combinations of styles, options and colors available on a certain new family sports car, for example, a computer expert came up with 25 million different versions of it for a buyer. And that is only a beginning. When automated electronic production reaches full potential, it will be just about as cheap to turn out a million differing objects as a million exact duplicates. The only limits on production and consumption will be the human imagination. (Marshall McLuhan and George B. Leonard, "The Future of Education: The Class of 1989," *Look*, 30 [February 21, 1967]: 23–25; French translation in McLuhan, *Mutations 1990* [Tours: Mame, 1969]).

Yet, until recently, nobody seems to have been aware of McLuhan's vaticinations. Charles Jencks has often referred to this eminent precedent in his recent works (Charles Jencks, *The Story of Post-Modernism* [Chichester: Wiley, 2011], 162). Philippe Morel has frequently cited McLuhan's 1967 article in his lectures, including at the conference *Digital Postmodernities* at the Yale School of Architecture in February 2014. McLuhan also precisely predicted the technical logic of the Web 2.0 ("At electric speeds the consumer becomes the producer and the public becomes a participant role player": Marshall McLuhan and Barrington Nevitt, *Take Today: The Executive as Dropout* [New York: Harcourt Brace Jovanovich, 1972], 4), a prediction that, as is often the case, went entirely unnoticed until well after it came true. In another quirky late-modern development, a 1969 essay by Reyner Banham celebrates multiple choice and user interaction as the alternative of the future to the "boredom" and alienation of industrial mass production. Banham gushes over the American hot rod movement (automotive enthusiasts who modified and customized their cars using industrial spare parts and catalog-based components), and, not surprisingly, in the same essay he disparages electronic technologies as a worrying, hostile development, and "computerisation" as an "invisible power system" that must be "diverted or disrupted." Reyner Banham, "Softer Hardware," *Ark* 44 (Summer, 1969): 2–11. *Ark* was the journal of the Royal College of Arts: number 44 was subtitled "is all about mass production/customisation," but the term "mass customization" itself was not yet used.

50. Carpo, "Ten Years of Folding," 14–19; Carpo, *The Alphabet and the Algorithm*, 83–93.

51. See in particular Peter Eisenman, "Visions Unfolding: Architecture in the Age of Electronic Media," *Domus* 734 (1992): 17–24; "The Affects of Singularity," in Andreas Papadakis, ed., "Theory + Experimentation," special issue (AD Profile 100) *Architectural Design* 62, nos. 11–12 (1992): 42–45; "Folding in Time: The Singularity of Rebstock," in *Architectural Design* 63, nos. 3–4 (1993): 38–42; Carpo, *The Digital Turn in Architecture*, 15–28.

52. The history of the notation, representation, and fabrication of free-form curves straddles disciplines and competencies: some parts of it pertain to the history of pre-industrial craft, others to the history of civil engineering and industrial manufacturing; others still to the history of mathematics and statistics, electronic computing, and software development; some are part of corporate industrial history and are protected by patents and by industrial and trade secrets. Reliable scholarly sources are few and far between; many—including corporate websites—list facts and dates that are patently contradictory or anecdotal. Sometimes the only source is *Wikipedia*, which, contrary to its terms of service, but faithful to its spirit, serves in this instance as an aggregator of oral traditions (mostly, one may infer, contributed by the protagonists themselves or people close to them). Among the sources most frequently used here, see in particular: Gerald Farin, "A History of Curves and Surfaces in CAGD," in Gerald Farin, Josef Hoschek and Myung-Soo Kim, *Handbook of Computer Aided Geometric Design* (Amsterdam and Oxford: Elsevier, 2002), 1–14; Georges Teyssot, with Olivier Jacques, "Faire parler les algorithms: Les nuages virtuels du Metropol Parasol (Séville) (Algorithms Can Talk: The Virtual Clouds of Metropol Parasol [Seville])," and Georges Teyssot, with Samuel Bernier-Lavigne, Pierre Côté, Olivier Jacques, and Dimitri Lebedev, "Des splines aux NURBS: aux origines du design paramétrique (From Splines to Nurbs: The Origins of Parametric Design)," *Le Visiteur, Revue Critique d'Architecture* 14 (2009): 101–121 and 122–123; Georges Teyssot, with Samuel Bernier-Lavigne, "Forme et Information. Chronique de l'Architecture Numérique," in *Action Architecture*, ed. Alain Guiheux (Paris: Éditions de la Villette, 2011), 49–87; Alastair Townsend, "On the Spline. A Brief History of the Computational Curve," Jonathan Anderson and Meg Jackson, eds., "Applied Geometries" issue, *International Journal of Interior Architecture + Spatial Design* 3 (2014): 48–60, accessed February 11, 2016, http://www.alatown.com/spline-history-architecture. See also: Malcolm Sabin, "Sculptured Surfaces Definitions: a Historical Survey," in *Techniques for Computer Graphic*, ed. David F. Rogers and Rae A. Earnshaw (New York: Springer, 1987), 285–338; David F. Rogers, *An Introduction to NURBS: With Historical Perspective* (San Francisco: Morgan Kaufmann, 2001); Philippe Morel, "From e-Factory to Ambient Factory (or What Comes After Research?)," in *GameSetAndMatch II: On Computer Games, Advanced Geometries and Digital Technologies*, transactions of the Conference, TU Delft, Netherlands, March 29– April 1, 2006, ed. Kas Oosterhuis and Lukas Feireiss (Rotterdam: Episode Publishers, 2006), 532–544; "A Few Precisions on Architecture and Mathematics," Mathematica Day, Henri Poincare Institute, Paris, January 31, 2004; "Géométrie polymorphe et jeux de langages formels: sur l'usage de la géométrie et des modèles dans l'architecture contemporaine," in *Modéliser & simuler: Epistémologies et pratiques de la modélisation et de la simulation, Tome 2*, ed. Franck Varenne, Marc Silberstein, Sébastien Dutreuil, and Philippe Huneman (Paris: Editions Matériologiques, 2014), 293–335; "Some Geometries:

Polymorphic Geometries and Formal Language Games," in Morel, *Five Essays on Computational Design, Mathematics and Production* (Sidney: Sidney University Press, 2008), 87–144. Other sources cited below.

53. For a full bibliography of Bézier's publication, see Christophe Rabut, "On Pierre Bézier's Life and Motivations," *Computer-Aided Design* 34, no. 7 (June 2002): 493–510. An abridged biography in David F. Roger's obituary, "Pierre Etienne Bézier, 1910–1999," in the same issue of *CAD*, 489–491; see also the editorial introduction to *Computer-Aided Design* 22, no. 9 (November 1990), a special issue honoring Pierre Bézier on his eightieth birthday; Pierre-Jean Laurent and Paul Sablonnière, "Pierre Bézier: An Engineer and a Mathematician," *Computer Aided Geometric Design* 18, no. 7 (September 2001): 609–617. Bézier's math was first published in Bézier, "Définition numérique des courbes et surfaces," *Automatisme* 11, no. 12 (1966): 625–632, and *Automatisme* 12, no. 1 (1967): 17–21; its application to CAD/CAM technology in Bézier, "How Renault Uses Numerical Control for Car Body Design and Tooling," Paper 6800010, *Society for Automotive Engineers* (Detroit, January 1968), Bézier, "Procédé UNISURF de Définition Numérique de Courbes Non Mathématiques," *Mécanique Electricité* 1968: 219, and various subsequent scholarly papers (1968, 1969, etc.; including, in 1971, in the *Proceedings of the Royal Society* (Mathematical and Physical Sciences 321, no. 1545 (1971): 207–218). The chronology of de Casteljau's findings is more controversial: "At Citroën, Paul de Casteljau defined the 'Bézier Curves' in a geometric and algorithmic way, by using the 'de Casteljau algorithm,' in late 1958 and 1959; he used Bernstein's polynomials soon afterward, much before Pierre Bézier. This has been presented in internal Citroën reports (such as: de Casteljau, P de F, SADUCSA manual, 1. Courbes à Pôles; 2. Surfaces à Pôles. Citroën [no date given]); was taught in the internal drawing school of Citroën from 1963; has been mentioned to the Institut de la Propriété Industrielle in March 1959 and June 1963; and has been kept secret by Citroën for a long time. Pierre Bézier was aware of this and used to fully recognize that his work was posterior to, though completely independent from, that of Paul de Casteljau." (Rabut, *CAD* 34, no. 7 [2002]: 508). According to Farin the first public mention of de Casteljau's algorithm, although not including a mention of the inventor, is in J. Krautter and S. Parizot, "Système d'aide à la définition et à l'usinage des surfaces de carrosserie," in P. Bézier, ed., "La commande Numérique" issue, *Journal de la SIA* 44 (1971): 581–586. "W. Boehm was the first to give de Casteljau recognition for his work in the research community. He found out about de Casteljau's technical reports and coined the term 'de Casteljau algorithm' in the late 1970s." (Farin, "A History of Curves and Surfaces in CAGD," see in particular section 1.3, "De Casteljau and Bézier," 4–6.) See Hanns Peter Bieri and Hartmut Prautzsch, "Preface," *Computer Aided Geometric Design* 16, no. 7 (August 1999), special issue dedicated to Paul de Faget de Casteljau: 579–581 (contains the text of the Laudatio read on the occasion of the conferral to de Casteljau of the Doctorate Honoris Causae by the University of Bern on December 6, 1997). De Casteljau's autobiographical note in the same issue is entertaining but does not provide a timeline (Paul de Faget de Casteljau, "My Time at Citroën," *CAGD* 16, no. 7 [1999]: 583–586).

54. P. Bézier, Letter to P. Rabut (1999), in Rabut, "On Pierre Bézier's Life and Motivations," *CAD* 34, no. 7 (2002): 504.

55. According to Farin, the term "spline" was first used in English in 1752 ("A History of Curves and Surfaces in CAGD," see in particular section 1.2, "Early Developments," 1–4, curiously citing as a source H. L. Duhamel du Monceau, *Élémens de l'Architecture Navale ou Traité Pratique de la Constructions des Vaisseux* [Paris 1752], which, however, being written in French, uses the term *courbes*). The *Oxford English Dictionary* cites a first occurrence in 1756 (from a book on beehives), deriving from East-Anglian dialects (*OED* entry last updated in 1989).

56. I. J. Schoenberg, "Contributions to the Problem of Approximation of Equidistant Data by Analytic Functions. Part A: On the Problem of Smoothing and Graduation. A First Class of Analytic Approximations Formulae," *Quarterly Journal of Applied Mathematics* 4, no. 1 (1946): 45–99; "Part B, On the Problem of Osculatory Interpolation. A Second Class of Analytic Approximation Formulae," *Quarterly Journal of Applied Mathematics* 4, no. 1 (1946): 112–137. On the timeline of B-Splines and NURBS development, see Farin, "A History of Curves and Surfaces in CAGD," in particular section 1.6, "B-spline Curves and NURBS," 9–10.

57. Farin, "A History of Curves and Surfaces in CAGD," section 1.8, "Subdivision Surfaces," 11–12. Chaikin's algorithm, the basis of today's software for surface subdivisions, was invented in 1974 (George M. Chaikin, "An Algorithm for High Speed Curve Generation," *Computer Graphic and Image Processing* 3, no. 4 [December 1974]: 346–349. George M. Chaikin, 1944–2007, first trained as an architect at The Cooper Union, then as a computer scientist, was an artist and educator.) Chaikin's algorithm's begot Doo and Sabin's, and Catmull and Clark's (both in 1978): see E. Catmull and J. Clark, "Recursively Generated B-Spline Surfaces on Arbitrary Topological Meshes," *Computer-Aided Design* 10, no. 6 (November 1978): 350–355; Daniel Doo and Malcolm Sabin, "Behavior of Recursive Division Surfaces Near Extraordinary Points," *CAD* 10, no. 6 (1978): 356–360. Philippe Morel has pointed out that a precedent for these and other recursive methods can be found in a small recreational work by the Swiss mathematician Georges de Rahm, "Un peu de mathématiques à propos d'une courbe plane," *Elemente der Mathematik, Revue de mathématiques élémentaires*, II, 4 (July 15, 1947): 73–88 (oral communication). Today's best-selling subdivision-based animation software, Maya, is hugely influential in the design community, and in schools of architecture around the world. For a vivid example on the use of Maya in the design of an icon of contemporary "globular" design (the Metropol Parasol in Seville, designed by Jürgen Mayer H., Berlin; competition 2004; inauguration 2011) see Teyssot, "Algorithms Can Talk," *Le Visiteur* 14 (2009): 206–212; for a timeline of the development of subdivisions algorithms, 210 and footnotes. Maya was first released in February 1998; it was developed by the company Alias/Wavefront (owned by Silicon Graphics as of 1995; formerly Alias Systems Corporation, formerly Alias Research, founded 1983; bought by Autodesk Inc., 2006): see the *Wikipedia* entries for Autodesk Maya and Alias System Corporation. The Autodesk corporate website does not provide any information on the history and development of the software.

58. P. Bézier, letter to C. Rabut (1999), "Les Méthodes de la Carrosserie in 1960," in Rabut, "On Pierre Bézier's Life and Motivations," *CAD* 34, no. 7 (2002): 495; Paul de Faget de Casteljau, "My Time at Citroën," *CAGD* 16, no. 7 (1999): 583–586.

59. De Casteljeau, "My Time at Citroën," 583.

60. Pierre-Jean Laurent, Paul Sablonnière, "Pierre Bézier," *CAGD* 18, no. 7 (2001): 610, 614: Bézier's team had a CAE (Compagnie d'Automatisme

Électronique) 530 computer (8Kb RAM), bought second-hand; "in spite of so crude an equipment, relatively complicated curves could be obtained: in order to demonstrate to financial backers the proficiency of his prototype, Bézier had the signature of the head cashier of the Banque de France, as found on any French bank note, retraced in big scale!" De Casteljau cites an Olivetti Tectractys (de Casteljau, "My Time at Citroën," 584), which must be a practical joke, as the Tectractys was an electromechanical calculator of the same class as those I described above (section 2.1).

61. De Casteljau, "My Time at Citroën," 585. On the development of UNISURF at Renault, see Pierre-Jean Laurent, Paul Sablonnière, "Pierre Bézier," 614.
62. Farin, "A History of Curves and Surfaces in CAGD," section 1.6. According to Farin, the first systematic treatment of NURBS goes back to K. Versprille's PhD thesis in 1975. According to Townsend, "On the Spline. A Brief History of the Computational Curve," *International Journal of Interior Architecture + Spatial Design* 3 (2014), the term NURBS was introduced by Boeing in 1981 (no source given). Developments at Boeing are cited by Farin ("A History of Curves and Surfaces in CAGD") but by none of the authors contributing autobiographical notes to David F. Rogers, *An Introduction to NURBS: With Historical Perspective*: Robin Forrest, Rich Riesenfeld, Elaine Cohen, Tom Lyche, Lewis Knapp, Ken Versprille, David Rogers.
63. David Rogers, "Biography: Pierre Bézier," in Rogers, *An Introduction to NURBS*, 36.
64. Robin Forrest, "Bézier Curves," in Rogers, *An Introduction to NURBS*, 13.
65. Interviews with Frank Gehry, Tensho Takemori, Rick Smith, in Greg Lynn, ed., *Archaeology of the Digital*, Catalogue of the Exhibition held at the Canadian Centre for Architecture, Montréal, Québec, May 2013 (Montréal: Sternberg Press, 2013): 25, 32, 38–39. Gehry had started to model the Barcelona Fish using Alias (an early animation software); Gehry's team was referred to IBM, then to Dassault by Bill Mitchell, then dean of the MIT School of Architecture (ibid., 39). The idea was that the design of a big sculpture in the shape of a fish would pose problems of streamlining similar to those tackled in aircraft design, an intuition that appears more than warranted by the technical and etymologic origin of splines in shipyard technology. Dassault's corporate websites state that CATIA was launched in 1981 (and adopted, among others,

by Boeing in 1984): accessed February 11, 2016, http://www.3ds.com/about-3ds/history/1981-1997. CATIA may have been developed and used internally by the Dassault aircraft company (then called Avions Marcel Dassault-Breguet Aviation) from 1977 to 1981, with the name of CATI, before the software was renamed CATIA to be marketed by the newly created subsidiary Dassault Systèmes ("CATIA" entry, *Wikipedia*).

66. Autodesk corporate website, accessed February 11, 2016, http://usa.autodesk.com/adsk/servlet/item?id=12012348&siteID=123112.

67. Timeline from the website of the software's publisher, AutoDesSys, accessed February 11, 2016, http://www.formz.com/home/aboutus.html. On the development of Form-Z by Chris Yessios and others at Ohio State University in the late 1980s, and the involvement of Peter Eisenman in the early phases of the project, see Greg Lynn, ed., *Archaeology of the Digital*, 55. "Chris Yessios would also claim that I had asked for something to model things in 3-D, and that he developed FormZ for our needs. We had special rights to FormZ for years. We were the guinea pigs." According to Lynn, Eisenman might have used an early version of FormZ, called Archimodos, for his Frankfurt Biozentrum project, submitted in 1987 (ibid., 57). Eisenman denies that (16). See also Pierluigi Serraino, *History of Form*Z* (Basel: Birkhäuser, 2002; and Turin: Testo e Immagine, 2002); and Luca Galofaro, *Digital Eisenman. An Office of the Electronic Era* (Basel: Birkhäuser, 1999). On the role of spline-based modeling software in the history of digitally intelligent architecture, see Malcolm McCullogh, "20 Years of Scripted Space," in Mike Silver, ed., "Programming Cultures," special issue (AD Profile 182), *Architectural Design* 76, no. 4 (2006): 12–15. Rhino version 1.0 was released in October 1998. For a Rhino timeline, see the website of the software development company, McNeel, http://wiki.mcneel.com/rhino/rhinohistory (accessed March 24, 2016).

68. It should be noted that not all that is digitally notated must be digitally produced—i.e., fabricated or otherwise physically materialized using discrete, number-based processes from start to finish. Images on today's computer screens are rasters of discrete pixels, and so is almost everything printed on paper using laser or ink-jet peripherals, for example. But many may remember the drafting plotters of the last century, where drawings were made, line by line, by the mechanical movements of a pen. Such machines could draw an arc of a circle by pivoting an arm around a center, thus in a sense imitating the gesture of the hand of an artisan, instead of calculating and then printing out a raster of very small dots or pixels, as today's printers do. Mechanical plotters need very little vectorial information to do so, whereas today's rasterized printouts use plenty of data, and need a different bitmap for each scale or size of operations; to the limit, at maximum resolution the number of pixels needed to rasterize a circle is infinite, whereas the equation (or the program for the movement of the mechanical arm of the plotter) needed to draw that circle is always the same: the drafting plotter was still a small-data tool, a data-compression technology using mathematics to translate a long list of points into a short line of script. Drafting plotters have almost entirely disappeared, but many similarly analog strategies persist in today's 3-D robotic fabrication, or whenever robotic arms are tasked to perform gestural operations that are digitally calculated but mechanically operated.

69. Bernard Cache, "Objectile: The Pursuit of Philosophy by Other Means," in "Hypersurface Architecture II," ed. Stephen Perella, special issue (AD Profile 141), *Architectural Design* 69, nos. 9–10 (1999): 67 ("mathematics has effectively become an object of manufacture"); for centuries, architects have been measuring their drawings with algebra, but now, "CAD software enables architects to draw and sketch using calculus." Greg Lynn, *Animate Form* (New York: Princeton Architectural Press, 1999), 16–18.

70. One of the best accounts to date of Gehry's early digital processes is in Alex Marshall, "How to Make a Frank Gehry Building," *New York Times Magazine*, April 8, 2001. See also Coosjie Van Bruggen, *Frank O. Gehry, Guggenheim Museum Bilbao* (New York: Guggenheim Museum, 1997), 135–140; Bruce Lindsey, *Digital Gehry: Material Resistance, Digital Construction* (Basel: Birkhäuser, 2001), 42–47, 65–69, and Alberto Sdegno, "E-architecture, L'architettura nell'epoca del computer," *Casabella* 691 (2001): 58–67.

71. The term was forcefully reinstated in architectural discourse by Patrik Schumacher, and thanks in particular to the reach and popularity of his recent writings, parametricism is now often used as a synonym for digitally intelligent architecture in general. Schumacher first used the term in a lecture at the Smart Geometry conference in Munich in 2007, and in print in two contributions to the catalog of the 11th Venice Architecture Biennale the following year, "Parametricism as Style—Parametricist Manifesto" and "Experimentation Within a Long Wave of Innovation," in *Out There: Architecture Beyond Building: Catalogue of the 11th International Biennale d'Architettura, Venice 2008*, 5 vols., ed. Aaron Betsky (Venice: Marsilio, 2008), esp. vols. 3 and 5. See also Patrik Schumacher, "Parametricism: A New Global Style for Architecture and Urban Design," in Neil Leach, ed., "Digital Cities," special issue (AD Profile 200), *Architectural Design* 79, no. 4 (2009): 14–23. In design literature, parametricism is often meant to refer to the generic notation of a curve, which, for example, when planar, is written as a function of two variables and several coefficients (as in: $y = ax^2 + bx + c$). The variation of the coefficients (a,b,c) between certain limits generates a variable family of curves (in that instance, parabolas). This notion of parametricism in design culture may be traced back to Gilles Deleuze's and Bernard Cache's definition of the *objectile* as a generic (parametric) object (see note 49). In mathematics, however, a parametric function more specifically defines the translation of an explicit function (as the one written above) into a system of equations rewritten as functions of an external parameter. Most of the mathematics of free-form curves and surfaces use parametric, not explicit notations of functions, as these are more convenient to describe curves as functions of angles, for example. As a consequence, "parametricism" in design can also more specifically refer to the use of software based on splines, B-Splines, NURBS, Bézier curves, and subdivisions. Schumacher may have had both meanings in mind when he relaunched the term in 2007–08. See the online version of his *Parametricist Manifesto* of 2008 (accessed February 11, 2016, http://www.patrikschumacher.com/Texts/Parametricism%20as%20Style .htm): "Contemporary avant-garde architecture is addressing the demand for an increased level of articulated complexity by means of retooling its methods on the basis of parametric design systems. The contemporary architectural style that has achieved pervasive hegemony within the contemporary architectural avant-garde can be best understood as a research programme based upon the parametric paradigm. We propose to call this style: *Parametricism*." In today's design parlance the terms may refer both to the use of parametric software for the design of free-form curves and surfaces (NURBS, splines, etc.), and to the general variability inherent in all parametric notations, regardless of form and style.

72. But see note 26.

73. IBM's Selectric typewriter first introduced the "typeball," or golf ball in 1961 (all websites accessed March 24, 2016, http://www-03.ibm.com/ibm/history/ibm100/us/en/icons/selectric). Xerox developed the daisy wheel printing technology as of 1972–73. On the development of the daisy wheel printer, see the *Wikipedia* entry (https://en.wikipedia.org/wiki/Daisy_wheel_printing); the Xerox Corporation website does not refer to the development of the daisy wheel printer at all (http://www.xerox.co.uk/about-xerox/history-timeline/engb.html).

74. Laser printing was developed by Xerox's Palo Alto Research Center (PARC) in 1971–72; see a brief timeline on the PARC website (http://www.parc.com/about). The first laser printers aimed at the desktop market were the HP Laserjet (1984; http://www.hp.com/hpinfo/abouthp/histnfacts/museum/imagingprinting/0018) and the Apple LaserWriter (1985). On the development of the Apple LaserWriter, see the *Wikipedia* entry (https://en.wikipedia.org/wiki/LaserWriter) and http://www.macworld.com/article/1150845/laserwriter.html, accessed Februrary 11, 2016. The Apple corporate website does not provide an official timeline.

75. The Adobe company was founded in 1982. In 1985 the Apple Laser Writer was the first printer to ship with a built-in PostScript interpreter (the "interpreter" was needed to rasterize the PostScript files sent out by the computer). Timeline from the Adobe corporate website; see in particular, accessed February 11, 2016, http://blogs.adobe.com/typblography/files/typblography/TT%20PS%20OpenType.pdf. See also Townsend, "On the Spline," *International Journal of Interior Architecture + Spatial Design* 3 (2014): 52.

76. EZCT Architecture & Design Research, Philippe Morel with Hatem Hamda and Marc Schoenauer, *Studies on Optimization: Computational Chair Design Using Genetic Algorithms*, 2004. First presented at Archilab 2004 in Orléans (France) and published in Philippe Morel, "Computational Intelligence: The Grid as a Post-Human Network," in Christopher Hight and Chris Perry, eds., "Collective Intelligence in Design," special issue [AD Profile 183],

Architectural Design 76, no. 5 (2006): 100–103. A prototype and drawings of the "Bolivar" model are in the Centre Pompidou Architecture Collection.

77. Daniel Widrig, *Degenerate Chair* (2012); Jenny Sabin, *eSkin* (2007–13); Ruy-Klein, *Klex* (2008); Alisa Andrasek and Jose Sanchez, *Bloom Games* (2012); Andrew Kudless, *Chrysalis III* (2012); Marcos Cruz and Marjan Colletti, *Robot Foaming* (2013) in Marie-Ange Brayer and Frédéric Migayrou, eds., *Naturaliser l'architecture (Naturalizing Architecture)*, Catalogue of the Exhibitions ArchiLab 2013, Orléans, FRAC, Fonds Régional d'Art Contemporain, September 14, 2013–February 2, 2014 (Orléans: Éditions HYX, 2013): 91, 143, 149, 207, 219, 267.

78. See note 34.

79. See note 40.

80. See note 68, on the analog implementation of digital notations, and other exceptions: not all that is digitally designed must be physically materialized using discrete, number-based processes from start to finish.

81. Before the development of electronic computers, NC (Numerical Control) milling machines were driven by electromechanical calculators, punched cards, or punched tapes. The first prototypes were developed by John T. Parsons (1913–2007, engineer, industrialist) for the US Air Force, then by the Servomechanisms Laboratory at MIT in 1949–52. As in the case of the industrial history of Bézier's curves ten years later, Parsons's research was originally triggered by the need to mass-produce splines (in this instance, the airfoils of helicopter blades); for his contributions to the field of numerical control John T. Parsons was the first recipient of the Numerical Control Society Jacquard Award in 1968. The first commercial NC unit was marketed in 1954–55 by the Bendix Corporation (James Benes, "Microprocessor Magic," *American Machinist* 140, no. 8 [August 1996]: 124–132). For a slightly different timeline, see William Pease, "An Automatic Machine Tool," *Scientific American* 187, no. 3 (1952): 101–115, and the corporate website of CMS North America, accessed February 11, 2016, http://www.cmsna.com/blog/2013/01/history-of-cnc-machining-how-the-cnc-concept-was-born. NC machines started to include transistors in 1960 and integrated circuits in 1967 (Benes, "Microprocessor Magic," 124); it is not known when the expression "CNC" (Computer Numerical Control) was first introduced. As late as 1996, however, engineers appeared to consider CNC milling machines as tools for batch-type manufacturing—that is, for the mass production of small series, where machines and production lines have to be reconfigured to switch from one batch of products to the next (ibid., 131). While batch-type manufacturing may be related to some notion of product diversification, the idea to use CAD/CAM technologies for what today we call digital mass customization did not originate from within the technical professions—nor was it intuited by postmodern sociologists or by marketing gurus. Digital mass customization was invented by the first generation of digitally intelligent designers.

82. T. Rowe Price, "Infographic: A Brief History of 3D Printing," *Connections*, May 2012, accessed February 11, 2016, http://individual.troweprice.com/staticFiles/Retail/Shared/PDFs/3D_Printing_Infographic_FINAL.pdf. See chapter 3, note 46.

83. See chapter 3, note 48.

84. The argument only applies if we exclude the authorial cost of writing each different page in the first example, and of designing each different volume in the second.

85. Michael Hansmeyer with Benjamin Dillenburger, *Grotto Prototype* (2012–13), now in the collection of the Frac Centre, Orléans; in Brayer and Migayrou, *Naturaliser l'architecture*, 76–77.

86. Hansmeyer and Dillenburger's grotto has a volume of 3.5 × 5 × 3.2 meters, and its solid mass is composed by 8 billion voxels. No human mind can notate so many different units one by one (and no human hand could fabricate all of them individually, regardless of the tools employed). While there are no physical limits to the number of voxels that a machine can print, the usual scientific and rational limits apply to the notation of an inordinate number of random discrete units: as the human mind can not easily enumerate an infinite list, human science tends to replace endless lists with more manageable, compressed entities (compression being obtained by way of indexation, sorting, generalization, or mathematical formulas, as previously discussed). Random aggregates of infinite sizes evidently exist in nature—indeed, this is the species under which nature is more often apprehended; which is why science strives to rationalize aggregates by converting them into simpler geometrical objects which can in turn be notated using equations and functions.

But in this instance, the complexity of the final product was achieved by *de facto* reversing the natural method of science: the authors started with simple geometrical objects (cubes, spheres, etc.), then used a sequence of iterative geometrical transformations to break them down into smaller and more fragmented units—ad infinitum, in theory; although the process stopped after eleven iterations and at a scale of resolution of 0.15 mm. Dillenburger and Hansmeyer call this method "mesh grammars," and they see it as a three-dimensional, generative upgrade of Stiny and Gips's shape grammars (first formulated in 1972). As they explain, the process is deterministic but, due to its complexity, its visual results cannot be anticipated by the designer; the design process is thus heuristic and fortuitous at all stages. Benjamin Dillenburger and Michael Hansmeyer, "The Resolution of Architecture in the Digital Age," in *CAAD Futures 2013*, CCIS 369, ed. J. Zhang and C. Sun (Berlin: Springer Verlag, 2013): 347–357; "Mesh Grammars: Procedural Articulation of Form," in R. Stouff et al., eds., *Open Systems: Proceedings of the 18th International Conference on Computer-Aided Architectural Design Research in Asia (CAADRIA 2013)*, 821–829. In theory, the algorithm could be stopped and reversed at all stages to get back to the simple volumes the designers started with, and, not surprisingly, the final product conspicuously bespeaks its rational nature and geometrical genesis. Yet, in spite of its regularity, the process embodies a logic that is the opposite of that of the human mind—and of human science: geometry is a powerful tool to streamline random events and to make (human) order out of (nature's) chaos; in this instance, geometry was used to create a posthuman chaos out of a human-made order. Dillenburger and Hansmeyer's mathematical method is, in a sense, spline making in reverse; and, indeed, in simpler cases and at lower resolutions, somewhat similar results can be obtained by reversing mainstream spline-modeling software.

87. Alberti's famous definition of ornament as supplement is set forth at the beginning of Book VI of *De Re Aedificatoria*: "Pulchritudo [est] certa cum ratione concinnitas universarum partium in eo, cuius sint, ita ut addi aut diminui aut immutari possit nihil, quin improbabilius reddatur. … Erit quidem ornamentum quasi subsidiaria quaedam lux pulchritudinis atque veluti complementum. Ex his patere arbitror, pulchirudinem quasi suum atque innatum toto essere perfusum corpore, quod pulchrum sit; ornamentum autem affícti et compacti naturam sapere magis quam innati." Alberti, *De Re Aedificatoria*, ed. and trans. Giovanni Orlandi (Milan: Il Polifilo, 1966), 6.2.5, 447–49; *On the Art of Building in Ten Books*, trans. Joseph Rykwert, Neil Leach, and Robert Tavernor (Cambridge, MA: MIT Press, 1988), 156: "Beauty is that reasoned harmony [*concinnitas*] of all the parts within a body, so that nothing may be added or taken away, or altered, but for the worse. ... Ornament may be defined as a form of auxiliary light and complement to beauty. From this it follows, I believe, that beauty is some inherent property, to be found suffused all through the body of that which may be called beautiful; whereas ornament, rather than being inherent, has the character of something attached or additional." See also Cosimo Bartoli's influential Italian translation: "La bellezza è un'certo che dibello, quasi come di se stesso proprio & naturale, diffuso per tutto il corpo bello, dove lo ornamento pare che sia un'certo che di appiccaticcio, & di attaccaticcio, piu tosto che naturale, o suo propio." *L'architettura di Leonbatista Alberti tradotta in lingua fiorentina da Cosimo Bartoli ... con la aggiunta de Disegni* (Florence: Torrentino, 1550), 163. Alberti goes on to devote most of books VII and VIII of his treatise to ornament, and this scrutiny includes most of Alberti's theory of the architectural orders (*columnationes*), in book VII. A freestanding, load-bearing column (in the Greek tradition) can hardly be seen as a dispensable addition to a building, but Roman orders (reliefs or half columns attached to a load bearing wall) can. In book IX Alberti returns to his theory of beauty (*concinnitas*, "the absolute and fundamental rule in Nature"), which is further defined as proportional beauty (based on numbers, proportions, and regular correspondence between parts): *De Re Aedificatoria*, 9.5.6; Orlandi, 816; Rykwert, 303. Proportional beauty (*concinnitas*) pervades and informs all beauty (*pulchritudo*) and ornament (*ornamentum*): 9.5.1, Orlandi, 81; Rykwert, 301.

88. Several publications bear witness to a recent surge of interest in the theory of ornament, suggesting a widespread awareness of the extent to which digital tools have altered the Western idea of ornament, both in the classical and in the modernist traditions. See, for example, Farshid Moussavi and Michael Kubo, *The Function of Ornament* (Barcelona: Actar, 2006); Alina Payne, *From Ornament to Object: Genealogies of Architectural Modernism* (New Haven: Yale

University Press, 2012); Antoine Picon, *Ornament: The Politics of Architecture and Subjectivity* (Chichester: Wiley, 2013).

89. See Mark Garcia, "Introduction," in Mark Garcia, ed., "Future Details of Architecture," *Architectural Design* 84, no. 4 (2014): 14–23.

90. Ibid., 22–23: "We could be at an event-horizon of the architectural detail in which the detail itself will exceed all of these precedents to become something almost entirely alien, other and unknown. ... The future details of architecture may well be biological-machinic hybrids or post-human architectural details."

91. Young and Ayata, Michael Young, *The Estranged Object* (Chicago: The Treatise Project c/o The Graham Foundation, 2015), esp. 45–51. Shklovsky first outlined his device (*priem ostraneniya*) in 1917 in the essay "Iskusstvo kak priem"—"Art as Device," republished in 1925 as the opening essay of his collection of writings *O Teorii Prozy*; published in English in Viktor Shklovsky, *Theory of Prose*, trans. Benjamin Sher (Champaign: Dalkey Archive Press, 1991), 1–14. Brecht's *Verfremdungseffekt* appeared for the first time in "Bemerkungen über die chinesische Schauspielkunst," published in English as "The Fourth Wall of China: An Essay on the Effect of Disillusion in the Chinese Theatre," *Life and Letters To-day* 15 (Winter 1936): 116–123.

92. The theory of the *Unzuhandenheit* was first outlined in *Being and Time* (1927). For a contemporary recast of Heidegger's tool analysis, see the recent work of Bruno Latour, in particular "Why Has Critique Run Out of Steam? From Matters of Fact to Matters of Concern," *Critical Inquiry* 30 (Winter 2004): 225–248; Graham Harman, "Heidegger on Objects and Things," in *Making Things Public: Atmospheres of Democracy*, ed. Bruno Latour and Peter Weibel (Karlsruhe, Germany: Center for Art and Media; 2005), 268–273; Graham Harman, "Technology, Objects and Things in Heidegger," *Cambridge Journal of Economics*, 34, no. 1 (2009): 17–25.

93. Jeff Wall, *Morning Cleaning*. Mies van der Rohe Foundation, Barcelona, 1999. See Young, *The Estranged Object*, 28–34.

94. Graham Harman, *Weird Realism: Lovecraft and Philosophy* (Winchester: Zero Books, 2012); Michel Houellebecq, *H. P. Lovecraft: contre le monde, contre la vie* (Monaco: Éditions du Rocher, 1991).

95. Karl Marx's theory of alienation (estrangement, *Entfremdung* in Marx's original) is outlined in his *Economic and Philosophic Manuscripts of 1844*: see in particular the *First Manuscript*, chapters XXIII–XXIV. Marx defines several modes of alienation induced by the economic separation between the worker and the ownership of the means of production, which characterizes capitalism, but also by the "physical and mental" separation between the worker and the technical logic of new modes of production, which characterizes the industrial system. The first argument is more often associated with Marx's critique of political economy, but Marx was no less eloquent in voicing the alienation of work brought about by technological change: "Political economy conceals the estrangement inherent in the nature of labor by not considering the direct relationship between the worker (labor) and production. It is true that labor produces for the rich wonderful things—but for the worker it produces privation. ... It [political economy] replaces labor by machines, but it throws one section of the workers back into barbarous types of labor and it turns the other section into a machine. It produces intelligence—but for the worker, stupidity, cretinism." *Economic and Philosophic Manuscripts of 1844*, trans. Martin Milligan (Moscow: Foreign Language Publishing House, 1959): xxiii–xxiv, first published in German in Marx and Engels, *Gesamtausgabe*, I, 3 (Moscow: Institute of Marxism-Leninism, 1932).

96. Todd Gannon, Graham Harman, David Ruy, and Tom Wiscombe, "The Object Turn: A Conversation," *Log* 33 (Winter 2015): 73–94, in particular 85: "DR: What would an object-oriented architecture look like? Designing something to look weird is not the answer," and 76: "TW: After a long period of focus on fluidity and connectivity, a new formal lexicon is in order. Chunks, joints, gaps, parts, interstices, contour, near-figure, misalignment, patchiness, low-res, nesting, embedding, interiority, and above all *mystery*, are terms that resonate for me." In the same issue Bryan E. Norwood hints at another interesting angle: Quentin Meillassoux's version of speculative realism was born as a critique of critical philosophy; in the history of philosophy, critical philosophy, or criticism, means the philosophy of Kant (as set forth in Kant's three *Critiques*: of *Pure Reason*, of *Practical Reason*, and of *Judgment*). But the term "postcritical" in architectural theory means a critique of the "critical project." Hence speculative realism, which is anti-critical in the Kantian sense of the term, must also be a great fit for postcritical architectural theory. See Bryan E. Norwood, "Metaphors for Nothing," *Log* 33 (Winter 2015): 117–118. Adding in the missing names, the above paralogism means that just like speculative realism is anti-Kantian, its architectural avatars are likely to be anti-Eisenmanian. The paralogism is perversely fallacious (as it builds on pure nominal accidents) but its conclusions are true.

97. On the history of "flat ontology" from Manuel de Landa to Levi Briant, see *Log* 33 (Winter 2015): 74. See Mark Foster Gage, "Killing Simplicity: Object-Oriented Philosophy in Architecture," *Log* 33 (Winter 2015): 103: "The true reality of an object is unknowable, but it does have perceivable qualities that Harman refers to as sensual ... OOO [object-oriented ontology] spans the divide, linking the perceivable with the unknowably complex, though not through a direct causality. Instead, OOO suggests that perceivable qualities can, through allusion, guide us into deeper realities." For a different view, see Peter Wolfendale, *Object-Oriented Philosophy: The Noumenon's New Clothes* (Falmouth: Urbanomic, 2014).

98. See Carpo, "Introduction," in *The Digital Turn in Architecture*, 10–12, and essays in that book by Charles Jencks, Michael Hensel, Achim Menges, and Michael Weinstock. On computational self-organization, see John Henry Holland, *Adaptation in Natural and Artificial Systems and Complexity: A Very Short Introduction* (Oxford: Oxford University Press, 2014). See also note 100.

99. *Mathematica* was first released in 1988. See Wolfram Research corporate website, accessed February 11, 2006, http://www.wolfram.com/company/background.html?source=nav.

100. Stephen Wolfram, *A New Kind of Science* (Champaign, IL: Wolfram Media, 2002). The concept of cellular automata derives from studies by John von Neumann and Stanislaw Ulam at the Los Alamos National Laboratory, from 1947 to 1953 (but Wolfram remarks that both still hoped to convert the results of their cellular automata experiments back to differential equations: ibid., 876). Cellular automata developed a cult following in the 1970s, following the publication of John Conway's *Game of Life* (a mathematical game, and soon afterward a solitaire videogame; first published in *Scientific American* in October 1970). The game resonated with early postmodern theories of randomness, indeterminacy, and complexity, and with coeval interests in adaptive systems and genetic or evolutionary algorithms (Holland, *Adaptation in Natural and Artificial Systems*). Wolfram published his first paper on cellular automata in 1983, but he claims that he started nurturing the same concepts in 1972, at the age of twelve. Wolfram, *A New Kind of Science*, 17.

CHAPTER 3

1. The American Standard Code for Information Interchange (ASCII), derived from the Morse code, originally provided a register of 128 slots, each associated with a decimal and binary integer. Most personal computers used from the start an 8-bit (2^8) ASCII extension—that is, a register of 256 characters, which is more than enough to cover an extended keyboard in lowercase and uppercase, plenty of diacritical signs, regional variations, and more. As one byte or octet of information may encode a decimal integer between 0 and 255, (1 byte = 8 bits = a sequence of 8 "0"s or "1"s, = 2^8 = 256 combinations), it is common today to count each alphanumerical character as 1 byte.
2. Alberti's definition of the recession in space of the vanishing point (which he calls the "central point" or "centricus punctus") is famously "to an almost infinite distance" ("quemadmodum paene usque ad infinitam distantiam"): Leon Battista Alberti, *De Pictura*, I.19, in Alberti, *On Painting and On Sculpture: The Latin Texts of De Pictura and De Statua*, ed. and trans. Cecil Grayson (London: Phaidon, 1972), 54–55, and in the Italian version, also Alberti's, "quasi persino in infinito": Alberti, *Opere Volgari*, 3 vols., ed. Cecil Grayson (Bari: Laterza, 1973), 3:36–37.
3. In fact, geographic information systems are getting there (using photogrammetric surveys conducted from aerial or elevated vantage points).
4. *Isidori Hispalensis episcopi Etymologiarum sive Originum libri XX*, ed. W. M. Lindsay (London: Oxford University Press, 1985; first published Oxford: Clarendon Press, 1911), XIX:16.1: "Pictura autem dicta quasi fictura; est enim imago ficta, non veritas. Hinc et fucata, id est ficto quodam colore inlita, nihil fidei et veritatis habentia."
5. William Ivins famously attributed to the mechanical reproduction of images a central role in the birth and development of modern science and technology: Without the "exact repetition of pictorial statements," he claims, "most of our modern highly developed technologies could not exist. Without them we

could have neither the tools we require nor the data about which we think." William M. Ivins Jr, *Prints and Visual Communication* (London: Routledge & Kegan Paul Limited, 1953), 3, 160. On the role of illustrated books in print for the dissemination of technical knowledge, see Elizabeth L. Eisenstein, *The Printing Revolution in Early Modern Europe* (Cambridge: Cambridge University Press, 1983), 185–252, and Mario Carpo, *Architecture in the Age of Printing* (Cambridge, MA: MIT Press, 2001), with further bibliography.

6. See the famous (and controversial) notion of a "non-visual" form of imitation in the Middle Ages in Richard Krautheimer, "Introduction to an 'Iconography of Mediaeval Architecture,'" *Journal of the Warburg and Courtauld Institutes* 5 (1942): 1–33, esp. 17–20. Reprinted in *Studies in Early Christian, Medieval, and Renaissance Art* (New York: New York University Press, 1969), 115–151, in particular 117–127, n82–86. The quote is from Richard Krautheimer and Trude Krautheimer-Hess, *Lorenzo Ghiberti* (Princeton: Princeton University Press, 1956), 294 ("To Petrarch it mattered little whether or not a site was commemorated by a monument, or merely haunted by memories. His approach was entirely literary, almost emphatically non-visual"). See also Françoise Choay, *L'allégorie du patrimoine* (Paris: Seuil, 1992), 39, n31.

7. Projections "project" a point in space from a center of projection (in Alberti's theory, that is one human eye) onto a picture plane, where that projection leaves a trace. All the points on the same visual beam, line, or ray (the line connecting the eye with a point being seen) being aligned, they intersect the picture plane in the same point, hence they will translate into a single point of the perspectival projection.

8. The parallel between the discovery of perspective and the invention of print with moveable type was first suggested by Vasari in the revised, 1568 edition of his *Lives*, where Vasari added a paragraph celebrating Alberti's discovery of an "instrument" for painting, capable of enlarging and diminishing figures in perspective (Giorgio Vasari, *Le Vite ... nelle redazioni del 1550 e 1568*, ed. Rosanna Bettarini and Paola Barocchi [Firenze: Sansoni, 1971], 3: 286). Vasari, who does not mention Alberti at all with regard to the invention of geometrical perspective, dates this discovery to 1457, and he assumes it to have been exactly coeval with Gutenberg's invention of print with moveable type. In 1830 Quatrèmere de Quincy brilliantly elaborated upon this Vasarian

passage, outlining a cultural and technical analogy between the discoveries of "optics" and the invention of print in the fifteenth century ("l'imprimerie multiplie les jouissances des ouvrages de l'esprit, l'optique celle des ouvrages de l'art et de la nature ... l'optique établit dans le monde une sorte de communication qui fait pour les choses, ce que l'imprimerie fait pour les idées." [Antoine-Chrysostome Quatremère de Quincy, *Histoire de la vie et des ouvrages des plus célèbres architectes, du XIe siècle jusqu'à la fin du XVIIIe, accompagnée de la vue du plus remarquable édifice de chacun d'eux* (Paris: Jules Renouard, 1830), 1: 80–96, 85.]). Friedrich Kittler, assuming that Vasari may have been attributing to Alberti the use of a camera obscura, concluded that "the camera obscura ... created reproductions of the world as free of copying errors as otherwise only Gutenberg's printed books were. ... Two simultaneous technologies had appeared, poised to eliminate the disturbances of the human hand from texts and from images" (Friedrich A. Kittler, "Perspective and the Book," *Grey Room* 5 [2001]: 38–53, 43, first published in German as "Buch und Perspektive," in *Perspektiven der Buch und Kommunikationskultur*, ed. Joachim Knape [Tübingen: Hermann-Arndt Riethmüller, 2000], 19–31). Vasari's passage may more likely have been referring to Alberti's "veil" (another image-making technology which Alberti describes in the second book of *De Pictura* as a more general alternative to the geometrical construction), but Kittler's conclusions equally apply to perspective as such, regardless of any mechanical or optical expedient meant to facilitate the task of the painter: Alberti's perspectival construction, when all its geometrical parameters are given, is a mathematical operation, identically replicable ad infinitum, and always leading to the same results, regardless of who carries it out. See Carpo, "Alberti's Media Lab," in Mario Carpo and Frédérique Lemerle, eds., *Perspective, Projections and Design* (London: Routledge, 2007), 47–63, esp. 52–55, 60–61; Mario Carpo, *The Alphabet and the Algorithm* (Cambridge, MA: MIT Press, 2011); esp. 2.3, "Windows," 58–61.

9. On modern visuality versus classical tactility the best source, in spite of its well-known quirkiness and some generalizations that would not easily pass muster today, is still William M. Ivins Jr., *Arts & Geometry: A Study in Space Intuitions* (Cambridge, MA: Harvard University Press, 1946).

10. Alberti, *De Pictura*, II.26. See also *On Painting and On Sculpture*, 61–63.

11. Pliny, *Natural History, Vol. IX: Books 33–35*, trans. H. Rackhman (Cambridge, MA: Harvard University Press, 1952; Loeb Classical Library 394), esp. 35.5, 270–273 (on the origins of painting from tracing lines around human shadows) and 35.43, 370–373 (on sculpture: Butades of Sicyon, or Corinth, a potter, saw a drawing his daughter had made of his lover by tracing the shadow of his profile thrown upon the wall by the light of a lamp. Butades then filled in the contour with clay, and made a terracotta bas-relief). Quintilian, *The Orator's Education, Books 9–10*, ed. and trans. Donald A. Russel (Cambridge, MA: Harvard University Press, 2001; Loeb Classical Library 127), esp. 10.2.7, 324 (the first painters only traced lines around human shadows projected by the sunlight; but from those beginnings, new discoveries ensued).
12. See Grayson on Alberti's sources (Alberti, *On Painting and On Sculpture*, 62). Alberti himself may have used mirrors to draw self-portraits (the anecdote is told by Vasari, albeit probably to suggest Alberti's limited talents as a painter: Vasari, *Vite*, 3: 289). On the importance of mirrors in Alberti's theory of painting see Grayson (Alberti, *On Painting and On Sculpture*, 112–113).
13. Ovid, *Metamorphoses, Volume I: Books 1–8*, trans. Frank Justus Miller; rev. G. P. Goold (Cambridge, MA: Harvard University Press; London: William Heinemann, 1916; 3rd ed., 1977; Loeb Classical Library 42) 3:403–510, 152–160. On Alberti's reference to Narcissus see Grayson (Alberti, *On Painting and On Sculpture*, 113); Daniel Arasse, "Le Destins de Narcisse," *Albertiana* IV (2001): 160–164 (with reference to Hans Belting, *Bild und Publikum im Mittelalter: Form und Funktion früher Bildtafeln der Passion* [Berlin: Mann, 1981]); Hubert Damisch, "L'inventeur de la peinture," *Albertiana* IV (2001): 165–187.
14. Alberti, *De Pictura*, II.27. See also Alberti, *On Painting and On Sculpture*, 64–65.
15. Starting with Alberti, who never distinguished between projected images and painting: his theory of what we now call geometrical perspective is set forth in a book he titled *On Painting* (*De Pictura*), where Alberti in fact never used the term *perspective*, as if all images were projections, and all painting perspectival.
16. Horace, *Ars Poetica*, 361, as quoted in Rensselaer W. Lee, *Ut Pictura Poesis. The Humanistic Theory of Painting* (New York: W. W. Norton & Co., 1967), 3, first published as an article in *The Art Bulletin* 22, no. 4 (1940): 197–269.

17. Erwin Panofsky, *Galileo as a Critic of the Arts* (The Hague: Martinus Nijhoff, 1954), 3. Benedetto Varchi, *Due lezzioni … nella prima delle quali si dichiara un Sonetto di M. Michelagnolo Buonarroti. Nella seconda si disputa quale sia piu nobile arte la Scultura, o la Pittura, con una lettera d'esso Michelagnolo, & piu altri Eccellentiss. Pittori, et Scultori, spora la Quistione sopradetta* (Florence: Torrentino, 1549). For a recent critical edition, see Benedetto Varchi, *Paragone. Rangstreit der Künste. Italienisch und Deutsch*, ed. and trans. Oskar Bätschmann and Tristan Weddingen (Darmstatd: WBG, 2013).

18. Codex Urbinas Latinus 1270, ff. 20 and following, cited from the critical edition in Paola Barocchi, *Scritti d'arte del Cinquecento*, vol. I (Milan-Naples: Riccardo Ricciardi, 1971), 474–488. Leonardo's text in the Codex Urbinas Latinus 1270 is dated by C. Pedretti ca. 1492 (Barocchi, *Scritti d'arte*, 474).

19. Ibid., 474.

20. Ibid., 478–479 and footnotes.

21. Lorenzo Lotto, *Portrait of a Goldsmith in Three Positions*, Vienna: Kunsthistorsches Museum, ca. 1530.

22. Anthony van Dyck, *Charles I*, Windsor Castle, Royal Collections, 1635–36.

23. Philippe de Champaigne, *Triple Portrait of Cardinal Richelieu*, London, National Gallery, 1642.

24. "La pittura … è di maraviglioso artificio, tutta di sottilissima speculazione" (Barocchi, *Scritti d'arte*, 481), "comendando … le cause delle sue dimostrazioni constrette dalla sua legge … abbraccia e ristringe in sé tutte le cose visibili, il che far non po la povertà della scultura, cioè li colori di tutte le cose, e loro diminuzioni" (484); "la prospettiva, investigazione e invenzione sottilissima delli studi matematici" (488); etc.

25. "Ed a me par bene, che l'una e l'altra [painting and sculpture] sia un'artificiosa imitazione di natura, ma non so già come possiate dir che più non sia imitato il vero, e quello proprio che fa la natura, in una figura di marmo o di bronzo, nella qual son le membra tutte tonde, formate e misurate come la natura le fa, che in una tavola, nella quale non si vede altro che le superficie, e quei colori che ingannano gli occhi. Né mi direte già, che più propinquo al vero non sia l'essere, che'l parere." Baldesar Castiglione, *Il libro del Cortegiano* (Venetia: Nelle case d' Aldo Romano & d' Andrea d' Asola, 1528), f. 24, cited from the critical edition in Barocchi, *Scritti d'arte*, 490. "And it seems to me that both

painting and sculpture are an artificious imitation of nature, but who would argue that a figure in marble or bronze, where all the members of the body are in full round, with the same shape and measures as in nature, is a worse imitation of truth, and of things as made by nature, than a painting on wood, where you can only see a surface, and colors that cheat the eye? Surely no one would not claim that resembling is closer to truth than being [the thing itself]." My translation.

26. Painting achieves "artificio maggiore in far quelle membra che scortano e diminuiscono a proporzion della vista con ragion di prospettiva, la qual *per forza di linee misurate* di colori di lumi e d'ombre vi mostra anco in una superficie di muro dritto il piano e il lontano." Castiglione, *Il libro del Cortegiano*, f.24v; cited from the critical edition in Barocchi, *Scritti d'arte*, 490. Emphasis mine. Painting achieves "a higher degree of art by making those members that shorten and diminish proportionally to sight following the rules of perspective, which, *by the power of measured lines* and of colors and lights and shadows, shows what is distant and far away even on the straight surface of a wall." My emphasis and translation.

27. Alberti, dedicatory letter of *De Pictura* to Brunelleschi ("Vederai tre libri: el primo, tutto matematico ..."). Alberti, *On Painting and On Sculpture,* 32, and on the manuscript tradition of this letter, 108. To be noted that at the beginning of *De Pictura* Alberti mitigates this argument, claiming he speaks as a painter, not as a pure mathematician, and for that reason he should be allowed to express himself in cruder terms ("pinguiore Minerva": *De Pictura* I.1; in Alberti, *On Painting and On Sculpture*, 37–38).

28. At the beginning of book I of *De Pictura* Alberti discusses briefly the nature of the visual rays ("there was considerable dispute among the ancients as to whether these rays emerge from the surface or from the eye ...") to conclude that, for the purposes of his new theory, such arguments are useless ("inutilis"), and can be set aside. Alberti, *De Pictura*, I.5 (in Alberti, *On Painting and On Sculpture*, 40–41).

29. "... delle tre dimensioni, due sole sono sottoposte all'occhio, cioè lunghezza e larghezza, che è la superficie, la quale da' Greci fu detta epifania, cioè periferia e circonferenza, perché delle cose che appariscono e si veggono, altro non si vede che la superficie, e la profondità non può dall'occhio esser compresa, perché la vista nostra non penetra dentro a' corpi opachi. Vede dunque l'occhio solamente il lungo e'l largo, ma non già il profondo, cioè la grossezza non mai. Non essendo dunque la profondità esposta alla vista, non potremo d'una statua comprendere altro che la lunghezza e la larghezza … senza profondità." "Only two of its three dimensions [of a statue] are actually exposed to the eye: length and width (which is the superficies, called *epifania* in Greek, that is to say, periphery or circumference). For, of the objects appearing and seen we see nothing but their superficies; their depth cannot be perceived by the eye because our vision does not penetrate opaque bodies. The eye, then, sees only length and width but never depth, and never thickness. Thus, since thickness is never exposed to view, nothing but length and width can be perceived by us in a statue." This is why, Galileo concludes, there is no point in imitating "la natura scultrice coll'istessa scultura; … artificiosissima imitazione sarà quella che rappresenta il rilievo nel suo contrario, che è il piano." "What will be so wonderful in imitating 'sculptress nature' by sculpture itself, in representing that which is relieved by the relief itself? Certainly nothing or very little, and the most artistic imitation will be that which represents relief on its opposite, which is the plane." Galileo, *Lettera* to Lodovico Cigoli of June 26, 1612, in Galileo Galilei, *Opere* (Florence: Barbera, 1890–1909), XI, 340–343. Italian cited from the critical edition in Barocchi, *Scritti d'arte*, 708–710. English translation from Panofsky, *Galileo as a Critic of the Arts*, 36–37. Panofsky takes due notice of Galileo's preference for planar over three-dimensional imitation (9), but does not comment on Galileo's arguments against the cognitive perception of depth.

30. Charles Wheatstone, "Contributions to the Physiology of Vision. Part the First. On some Remarkable, and hitherto unobserved, Phenomena of Binocular Vision," *Philosophical Transactions of the Royal Society of London* 128 (1838): 371–394. "Part the Second … (continued)," *Philosophical Transactions of the Royal Society of London* 142 (1852): 1–17. Online at http://rstl.royalsocietypublishing.org (accessed March 3, 2016).

31. Wheatstone, "Contributions. Part the First," 380.

32. Leonardo noted the difference between the images in each eye only to conclude that painters should ideally work from one eye, not two, to comply with the geometry of the perspectival construction, which admits only one center of projection. Ibid., 372. On alternative theories of monocular vision in Ptolemy (127–148 AD), Alhazen (1000 AD), Kepler (1611), Descartes (1637), Huygens (1667), and Newton (1704), see Robert A. Crone, "The History of Stereoscopy," *Documenta Ophtalmologica* 81 (1992): 1–16.

33. See, for example, Greenberg on cubism and collage: "There is no question but that Braque and Picasso were concerned, in their Cubism, with holding on to painting as an art of representation and illusion. But at first they were more crucially concerned, in and through their Cubism, with obtaining sculptural results by strictly nonsculptural means; that is, with finding for every aspect of three-dimensional vision an explicitly two-dimensional equivalent, regardless of how much verisimilitude might suffer in the process. Painting had to spell out, rather than pretend to deny, the physical fact that it was flat, even though at the same time it had to overcome this proclaimed flatness as an aesthetic fact and continue to report nature." Clement Greenberg, "Collage," in *Art and Culture. Critical Essays* (Boston: Beacon Press, 1961), 70–84, esp. 71.

34. Pietro Accolti, *Lo inganno de gl' occhi. Prospettiva Pratica* (Florence: Cecconcelli, 1625). See Filippo Camerota, "'The Eye of the Sun': Galileo and Pietro Accolti on Orthographic Projections," in Carpo and Lemerle, eds., *Perspective, Projections and Design*, 115–125, see esp. 123.

35. The reversal of the perspectival construction requires the objects in the picture to have some visible alignments, hence it may be more or less feasible according to the geometry of the objects shown in the picture.

36. Alberti, *De Re Aedificatoria*, ed. Giovanni Orlandi (Milano: Il Polifilo, 1966), 2.1.4; *On the Art of Building in Ten Books*, trans. Joseph Rykwert, Neil Leach, and Robert Tavernor (Cambridge, MA: MIT Press, 1988), 34. Alberti does not mention sections, which will be added, in a very similar context, in Raphael's "Letter to Leo X," ca.1519, chap. XVIII–XXI, Archivio Privato Castiglioni, Mantua. For a recent critical edition of the text, see Francesco Paolo di Teodoro, *Raffaello, Baldassar Castiglione e la* Lettera a Leone X (Bologna: Nuova Alfa Editoriale, 1994), 63–85. See Carpo, *The Alphabet and the Algorithm*, 134–135.

37. Gaspard Monge, *Géométrie Descriptive. Leçons données aux Ecoles Normales l'An 3 de la République* (Paris: Baudouin, VII [1798–99]).

38. See Peter Jeffrey Booker, *A History of Engineering Drawing* (London: Chatto and Windus, 1963), 114–128, 198–213; and Massimo Scolari, *Il Disegno Obliquo. Una storia dell'antiprospettiva* (Venice: Marsilio, 2005); published in English as *Oblique Drawing: A History of Anti-Perspective* (Cambridge, MA: MIT Press, 2012).

39. Alberti, *De Statua*, in Alberti, *On Painting and On Sculpture*, 117–143. *De Statua* was composed in Latin at some point between 1435 and 1472. Alberti used a similar number-based technology to scan and copy a map of Rome, (*Descriptio Urbis Romae*, in *Leon Battista Alberti's Delineation of the City of Rome*, ed. Mario Carpo and Francesco Furlan (Tempe, AZ: Center for Medieval and Renaissance Texts and Studies, 2007). See also Carpo, *The Alphabet and the Algorithm*, esp. 2.2, "Going Digital," 54–58.

40. The famous expressions are from Michelangelo's *Letter* to Benedetto Varchi (*Due lezzioni*, 155); but the distinction between additive and subtractive making is in fact another Albertian invention, first introduced in *De Statua*, 1.2 (Alberti, *On Painting and On Sculpture*, 120–121). Michelangelo famously thought that only the effort of taking matter away from a block of solid stone ("per forza di levare") was worthy of the name of sculpture; for him, additive sculpture was similar to painting.

41. Samuel F. B. Morse, *His Letters and Journals, edited and supplemented by his son Edward Lind Morse*, 2 vols. (Boston: Houghton Mifflin, 1914), I: 245. See the letter of August 22, 1823, with reference to his invention of a "machine for sculpture" (247) that would deliver "perfect copies of any model." Morse's hopes for a profitable commercial venture were thwarted when it appeared a patent for the machine would have infringed on one obtained by Thomas Blanchard in 1820. Blanchard's patent would likely be US Patent USoXoo3131, now partly online on the website of the United States Patent and Trademark Office, March 3, 2016, http://patft.uspto.gov. See Samuel Irenaues Prime, *The Life of Samuel F. B. Morse, LL.D., Inventor of the Electro-Magnetic Recording Telegraph* (New York: Appleton & Co, 1875), 127–130.

42. Morse, *His Letters and Journals*, II: 37–39; Samuel F. B. Morse, *Lectures on the Affinity of Painting with the Other Arts*, ed. Nicolai Cikovsky (Columbia: University of Missouri Press, 1983), 43.

43. Ibid., 139. Morse cites the English translation of Leonardo's *Treatise of Painting* printed in London by J. Taylor in 1802, which does not include any work of Alberti's. The first edition of Leonardo's *Trattato della pittura* was published in Paris in 1651 together with Bartoli's Italian translations of Alberti's *De Pictura* and *De Statua*. From Bartoli's Italian, Alberti's *De Statua* was translated into English first by John Evelyn in 1664 (in Evelyn's translation of Fréart de Chambray's *Parallèle de l'architecture antique avec la moderne*; reprinted in 1706–07 and in 1723), then by James Leoni in 1726 (reprinted in 1755).

44. US Patent 2,519, "Apparatus for Sculptors to be Employed in Copying Busts, etc.," now searchable under number US0000002519 on the website of the United States Patent and Trademark Office, March 3, 2016, http://patft.uspto.gov.

45. See chapter 2, note 70.

46. Accessed April 1, 2015, http://www.makerbot.com/blog/2012/01/09/introducing-the-makerbot-replicator. The MakerBot company eventually made some controversial marketing choices, and after a few years of rapid growth, it announced in April 2015 it would fire 20 percent of its staff. For a timeline of 3-D printing, see T. Rowe Price, "Infographic: A Brief History of 3D Printing," *Connections*, May 2012. Accessed February 11, 2016, http://individual.troweprice.com/staticFiles/Retail/Shared/PDFs/3D_Printing_Infographic_FINAL.pdf.

47. Michael Hansmeyer with Benjamin Dillenburger, *Grotto Prototype* (2012–13), in *Naturaliser l'architecture, Naturalizing Architecture*, Catalogue of the Exhibitions ArchiLab 2013, Orléans, FRAC, Fonds Régional d'Art Contemporain, September 14, 2013–February 2, 2014, ed. Marie-Ange Brayer and Frédéric Migayrou (Orléans: Éditions HYX, 2013): 76–77; *Digital Grotesque* (2013), on the architects' website, April 1, 2015, http://www.michael-hansmeyer.com/projects/digital_grotesque_info.html. See chapter 2, note 85.

48. For the text of the president's 2013 State of the Union address, accessed June 1, 2015, see http://www.whitehouse.gov/the-press-office/2013/02/12/

remarks-president-state-union-address. On the comments that followed the president's address, see Nick Bilton, "Disruptions: On the Fast Track to Routine 3-D Printing," *New York Times*, February 18, 2013. Accessed June 1, 2015, http://bits.blogs.nytimes.com/2013/02/17/disruptions-3-d-printing-is-on-the-fast-track. Around that time, 3-D printing also featured prominently in American popular culture (in US late night talk show *The Colbert Report* on June 8, 2011, for example, and, less than two weeks before President Obama's speech, in an episode of the American sitcom *The Big Bang Theory* on January 31, 2013).

49. The only source for the history of the device at the time of writing is *Wikipedia*; Microsoft's corporate website only describes the current version of the software and hardware, March 3, 2016, http://www.microsoft.com/en-us/kinectforwindows/meetkinect/about.aspx. See also note 50.

50. Jason Tanz, "Kinect Hackers are Changing the Future of Robotics," *Wired* 19, no. 7 (July 2011). Accessed June 1, 2015, http://www.wired.com/2011/06/mf_kinect; Rob Walker, "Freaks, Geeks and Microsoft. How Kinect Spawned a Commercial Ecosystem," *New York Times Sunday Magazine*, June 3, 2012. Accessed June 1, 2015, http://www.nytimes.com/2012/06/03/magazine/how-kinect-spawned-a-commercial-ecosystem.html.

51. Wheatstone, "Contributions. Part the First," 380.

52. Accessed March 3, 2016, http://www.123dapp.com/catch.

53. Accessed March 3, 2016, https://www.google.com/atap/project-tango.

54. Accessed March 3, 2016, http://www.factum-arte.com.

55. Accessed March 3, 2016, http://www.scanlab-ucl.co.uk, http://scanlabprojects.co.uk/visualisation.

56. The French company Photomaton, known for owning and operating thousands of automatic photo booths in public places, has recently launched a 3-D photo booth that, along with traditional ID pictures, construes a volumetric scan of the bust of the client; a statuette can then be 3-D printed off-site and shipped to the address provided. August 17, 2016, http://www.photomaton.fr/innovations/cabine_3d.

57. Accessed August 17, 2016, http://www.lytro.com/about.

58. See Facebook's Oculus Rift, Samsung's Gear VR, or Microsoft's Hololens.

59. William J. Mitchell, *The Reconfigured Eye. Visual Truth in the Post-Photographic Era* (Cambridge, MA: MIT Press, 1992). See Carpo, "The Photograph and the Blueprint. Notes on the End of Some Indices," in *Das Auge der Architecktur*, ed. Andreas Bayer (Munich: Wilhelm Fink Verlag, 2011), 467–482.
60. Jacopo Pontormo, letter dated 18 February [1546] in Varchi, *Due lezzioni*, 134. See Panofsky, *Galileo as a Critic of the Arts*, 9.

CHAPTER 4

A version of this essay was published as "Digital Style," *Log* 23 (Fall 2011): 41–52.

1. See chapter 2, note 71.
2. The term "Web 2.0" appears to have been introduced by the publisher and technologist Tim O'Reilly, and it rose to prominence after the first Web 2.0 Conference held by O'Reilly Media in San Francisco, October 5–7, 2004.
3. Mario Carpo, *The Alphabet and the Algorithm*, (Cambridge, MA: MIT Press, 2011), 123–128.
4. See in particular James Surowiecki, *The Wisdom of Crowds* (New York: Doubleday, 2004; repr. New York: Anchor, 2005), chapter 1, "The Wisdom of Crowds," 3–22; chapter 11, "Markets: Beauty Contests, Bowling Alleys, and Stock Prices," 224–258; and chapter 12, "Democracy: Dreams of the Common Good," 259–271.
5. See the afterword added to the 2005 edition of Surowiecki, *Wisdom of Crowds*, 273–282; and Howard Rheingold's pioneering *Smart Mobs: The Next Social Revolution* (Cambridge, MA: Perseus, 2002).
6. See the vast literature on the so-called Efficient-Market Hypothesis: if all market participants can interact with all others at all times (and digital information and communication technologies can get fairly close to that), market valuations always reflect all available information, hence market prices are always "right."
7. Trademarked by Google and named after Larry Page, cofounder of the company. On the engineering of Google's information retrieval, see chapter 2, note 26.
8. This famous statement appeared on Google's corporate website as late as June 27, 2011, but when this article was first written (August 2011), the chapter on

search relevance on Google's website had already been rewritten to take into account the customization of search results. See "PageRank," *Wikipedia*, accessed August 1, 2011, n. 4. At the time of revising this article (March 2016), the reference had also been removed from the *Wikipedia* entry, March 21, 2016, https://en.wikipedia.org/wiki/PageRank.

9. Early modern authorship rose in sync with print technologies, which favored a new format of identically reproduced, "authorized" textual versions protected and separated from the permanent drift of scribal copies. Not coincidentally, that notion of authorship is now being demoted by the new variability of digital media.

10. On the difference between "information products" and "physical products" in user-centered innovation and peer production, see Eric von Hippel, *Democratizing Innovation* (Cambridge, MA: MIT Press, 2005), esp. 1–6.

11. This idea was prefigured by Marshall McLuhan in 1972. See chapter 2, note 49.

12. A web search on the exact phrase "ways to get to the same result" returns thousands of hits from all sorts of software tutorials, often claiming that this heuristic approach makes learning more natural, spontaneous, or amusing.

13. Alexandre Koyré, "Du monde de l''à peu près' à l'univers de la précision," *Critique* 28 (1948): 806–823; reprinted in Koyré, *Études d'histoire de la pensée philosophique* (Paris: A. Colin, 1961), 311–329. Koyré's view of a clear-cut linear progression from ancient negligence to a modern quest for exactitude is often seen today as simplistic and West-centered.

14. For what an anecdote may be worth, I can refer to the case of the assistant to a very important person who has been curating said person's biographic entry in *Wikipedia* for the last ten years, and has developed over time a special literary skill: when editing or updating that entry, she redacts all new text in a deliberately fragmentary way, sometimes even introducing partially inconsistent or redundant information; the purpose being that her own authorial text should read just like a real *Wikipedia* entry should—that is, as if it had been written by many people of variable literary talent editing one another over time. Digital aggregation has already begot a recognizable literary style—the Wikipedic style of many hands.

15. Interactivity and variability are considered here only as attributes of the design process and as reflected in architectural notations, which are pure data. The material variability, interactivity, or responsiveness that may be built in some buildings, structures, or design pieces is not relevant to this discussion.

16. Carpo, *The Alphabet and the Algorithm*, 83–93, 123–129; "The Craftsman and the Curator," in Tala Gharagozlou, David Sadighian, and Ryan Welch, eds., "Domain," special issue, *Perspecta* 44 (2011): 86–91. See note 21 and chapter 2, notes 49, 71.

17. BIM technologies, as they took shape in the early 2000s, represent to date the most advanced—and, in many ways, daring—Web 2.0 experiment in architecture and design, and are likely the only significant example of transition from mass customization to mass collaboration in the design professions. There is no reliable historical timeline of the development of the concept of BIM, other than that provided by its protagonists: see, for example, the *Wikipedia* entry on "Building Information Modeling," accessed December 12, 2015, https://en.wikipedia.org/wiki/Building_information_modeling#BIM_origins_and_elements. The problem is compounded by the absence of a consensual definition of BIM, which makes a history of BIM almost coextensive with the history of computer-aided design, or even of computing itself. See, however, Chuck Eastman et al., eds., *BIM Handbook, A Guide to Building Information Modeling for Owners, Managers, Designers, Engineers, and Contractors* (Hoboken, NJ: Wiley, 2008), xi-xii; Peggy Deamer and Phillip G. Bernstein, eds., *Building (in) the Future: Recasting Labor in Architecture* (New York: Princeton Architectural Press, 2010); Richard Garber, ed., "Closing the Gap: Information Models in Contemporary Design Practice," special issue (AD Profile 198), *Architectural Design* 79, no. 2 (2009); Richard Garber, *BIM Design: Realizing the Creative Potential of Building Information Modeling* (Chichester: Wiley, 2014).

18. See Phillip Bernstein, "A Way Forward? Integrated Project Delivery," *Harvard Design Magazine* 32 (2010): 74–77; Deamer and Bernstein, *Building (in) the Future*, esp. Bernstein's essay, "Models for Practice: Past, Present, Future," 191–198; Carpo, "Forward," in Garber, *BIM Design*, 8–13.

19. In a strange and visionary book that is one of the foundations of the open-source movement, Eric S. Raymond calls the initiator and moderator of an open-source project a "wise leader" and a "benevolent dictator." Eric S. Raymond, *The Cathedral and the Bazaar: Musings on Linux and Open Source by an Accidental Revolutionary* (Beijing: O'Reilly Media, 1999), 101–102, 111. See also Carpo, "The Craftsman and the Curator."
20. On the role of digital tools in the design and construction of Eisenman's Aronoff Center for Design and Art, University of Cincinnati (1988–96), see David Gosling, "Peter Eisenman: Addition to the College of Design, Architecture, Art and Planning," *Architectural Design* 67, nos. 9–10 (September–October 1997): iii–xi; and in the same issue, Charles Jencks, "Nonlinear Architecture: New Science = New Architecture?," 7.
21. On one full-fledged application of digital parametricism that highlighted user interaction and participatory design, Bernard Cache's *Tables Projectives* (2004), see Carpo, *The Alphabet and the Algorithm*, 103–104 and 157, note 48.

CHAPTER 5

Versions of this chapter were previously published as "Micro-Managing Messiness," *AA Files* 67 (2103): 16–19; "Micro-managing Messiness: Pricing, and the Costs of a Digital Non-Standard Society," in James Andrachuk, Christos C. Bolos, Avi Forman, and Marcus A. Hooks, eds., "Money," special issue, *Perspecta* 47 (2014): 219–226 (with illustrations by Emily Orr).

1. Fixed prices were also needed for another of Boucicaut's innovations, the money-back guarantee. See Robert Phillips, "Why Are Prices Set the Way They Are?," in *The Oxford Handbook of Pricing Management*, ed. Özalp Özer and Robert Phillips (Oxford: Oxford University Press, 2012), 13–44; on fixed pricing see section 2.5, "Why do retailers sell at fixed prices?," 28–38; on Boucicaut see in particular section 2.5.3, "Why fixed pricing?," 33–35.
2. Ibid., 33–34.
3. Ibid., 29.
4. See Mario Carpo, *The Alphabet and the Algorithm* (Cambridge, MA: MIT Press, 2011), and "Introduction," in *The Digital Turn in Architecture, 1992–2012* (Chichester: Wiley, 2013), 8–14; Chris Anderson, *Makers: The New*

Industrial Revolution (New York: Random House, 2012), 81–98; chapter 2, notes 49, 71.

5. See chapter 4, note 6, and elsewhere.
6. The story is told in Dio Cassius's *Roman History, Volume IX: Books 71–80*, trans. Earnest Cary and Herbert B. Foster (Cambridge, MA: Harvard University Press, 1927; Loeb Classical Library 177), LXXIV:11, see esp. 142–143.
7. See chapter 4, notes 7, 8.
8. In 1999 Coca-Cola unsuccessfully tested a vending machine with prices that changed based on demand; more recently, outrage ensued when Amazon was suspected of charging different prices to different customers based on their past buying behavior. See Robert Phillips, *Pricing and Revenue Optimization* (Stanford: Stanford University Press, 2005), 302–303, 312.
9. The "e-hailing" upstart Uber was founded in 2009, and it started to expand internationally and to gain media attention in 2012. See Marcus Wohlsen, "Uber is Back for NYC Cabs as Taxi App Wars Escalate," accessed March 16, 2016, http://www.wired.com/2012/12/uber-flywheel-taxi-app-wars.
10. Jean-François Lyotard, *La condition postmoderne* (Paris: Les Éditions de Minuit, 1979), 31; published in English as *The Postmodern Condition: A Report on Knowledge*, trans. Geoff Bennington and Brian Massumi; foreword, Fredric Jameson (Minneapolis: University of Minnesota Press, 1984). Lyotard's original expression was "décomposition des grands Récits."
11. Hong Kong and Singapore are the best-known cases. The City of London, a medieval corporate body that has maintained many of its traditional privileges and franchises, is a less well-known example. Outside of the Western political tradition, some political and religious entities appear to be merging premodern tribalism and digital technologies to create territorial institutions alien and opposite to the modern nation state as it was crafted in the West during the industrial revolution. ISIL's caliphate is a case in point.

POSTFACE: 2016

1. On the notion of "favorable environment" in the history of technological innovations, see André Leroi-Gourhan, *Milieu et techniques* (Paris: Albin Michel, 1992), 373–377 (first published Paris: Albin Michel, 1945).

2. Rephrasing the celebrated topos by the Protestant theologian Matthäus Richter on the use of print to disseminate the words of Martin Luther (1566). See chapter 2, note 20.
3. See in particular Jeremy Rifkin, *The Zero Marginal Cost Society. The Internet of Things, The Collaborative Commons, and the Eclipse of Capitalism* (New York: Palgrave Macmillan, 2014).

INDEX